NATURAL HISTORY OF THE PHLOX FAMILY

VOL. I

SYSTEMATIC BOTANY

INTERNATIONAL SCHOLARS FORUM

A SERIES OF BOOKS BY AMERICAN SCHOLARS

SCIENCES

1

Polemonium eximium on Mount Dana, California.

NATURAL HISTORY
OF THE
PHLOX FAMILY

*

VOLUME 1

SYSTEMATIC BOTANY

by

VERNE GRANT

RANCHO SANTA ANA BOTANIC GARDEN,
AND CLAREMONT GRADUATE SCHOOL,
CLAREMONT, CALIFORNIA

Springer Science+Business Media, B.V.
1959

ISBN 978-94-017-5725-6   ISBN 978-94-017-6077-5 (eBook)
DOI 10.1007/978-94-017-6077-5

Copyright 1959 by Springer Science+Business Media Dordrecht
Originally published by Martinus Nijhoff, The Hague, Netherlands in 1959
Softcover reprint of the hardcover 1st edition 1959

To My Partner

ALVA GRANT

PREFACE

Believing that it is always best to study some special group, I have, after deliberation, taken up domestic pigeons.

Charles Darwin, 1859.

Within the territory of evolution studies ample niches lie open for inquiries projected on a limited scale and centered on a single group of organisms. Combined taxonomic, ecological and genetic studies of a single phyletic unit – genus or family – have made a large contribution to our emerging general theory of evolution. We have recently seen the publication of several such studies. Of particular value are *The Evolution of Gossypium* by Hutchinson, Silow and Stephens in 1947; *Darwin's Finches* by Lack, 1947; Babcock's *The Genus Crepis*, 1947; Manton's *Problems of Cytology and Evolution in the Pteridophyta*, 1950; Simpson's *Horses*, 1951; the work on the Tarweed Tribe embodied in Clausen's *Stages in the Evolution of Plant Species*, 1951; *Evolution in the Genus Drosophila*, by Patterson and Stone in 1952; Goodspeed's *The Genus Nicotiana*, 1954; and Ford's *Moths*, 1955.

It has been my goal to treat the Phlox family similarly as a model of evolution in a group of higher plants. The family is to be viewed as a model, not in the abstract sense of the theoretical and mathematical evolutionist, but rather in the sense of the naturalist – as a complex and far-flung production of nature. Three continents, but primarily western North America, are the physical setting in this model, and five tribes, eighteen generic lines, and approximately 316 species (most of these highly complex in themselves) comprise the dynamic elements in the system.

Before the biological information necessary for an understanding of evolution in the Phlox family can be effectively presented, or in some cases even effectively gathered, it is essential to lay the taxonomic groundwork. It is appropriate to begin our broad comparative study of the Phlox family with a taxonomic survey. This is followed by chapters dealing with chromosome numbers and karyotype evolution, phylogeny, and phytogeography including migrational history.

Research is in progress on a second volume dealing with the reproductive biology of the Phlox family and containing separate chapters on the life cycle; flower pollination; breeding systems; the genetic structure of populations, races and species; fertility relationships; and adaptive radiation in the reproductive system. The general plan was to incorporate the descriptive material into the present work and place the experimental findings in the sequel.

ACKNOWLEDGMENTS

In the course of preparing the present study I have received help from many colleagues. Dr. Edgar T. Wherry carefully reviewed Chapters 4 and 5 in manuscript, and I have benefitted greatly from his knowledgeable discussions. Mrs. Alva Grant contributed helpful suggestions concerning the infrageneric classification of *Gilia* and *Navarretia*. Dr. Sherwin Carlquist read Chapters 3 and 7, making a number of important suggestions. Discussions with Drs. Daniel I. Axelrod, Lyman Benson, Jay M. Savage, Barton H. Warnock and Miss Elizabeth Sprague concerning various aspects of Chapter 8 were most profitable. A discussion with Dr. G. Ledyard Stebbins concerning material in Chapter 1 was likewise very helpful. Drs. Philip A. Munz, Arthur Cronquist and Edgar T. Wherry helped on many bibliographical questions. I am indebted to Dr. Herbert L. Mason for several points of view expressed in Chapters 3 and 8.

Among botanists who have aided in the collection of seeds of Polemoniaceae in various parts of the world are: A. Beetle (Wyoming and Argentina); D. Dunn (Montana); A. Garaventa (Chile); J. Hunziker (Argentina); H. Johnson (Peru); E. Landolt (Switzerland); P. A. Munz (California); M. Reiche (Peru); M. Ricardi (Chile); J. Rzedowski (Mexico); R. Shaw (Utah); A. Soriano (Argentina); B. Turner (Texas); and Wm. A. Weber (Colorado). The seed collections contributed by these individuals enabled me to study living plants of a wider variety of Polemoniaceae than would otherwise have been possible.

Valuable technical assistance in the cytological phases of the investigation was provided by Mr. Howard Latimer. The geographical localities of certain Polemoniums previously determined for chromosome number were kindly furnished by Drs. C. A. Berger, W. S. Flory, and A. Nygren. Similar data for *Phlox* were provided by Drs. J. R. Meyer, E. T. Wherry, and O. E. White. For permission to quote their unpublished chromosome count of *Ipomopsis gossypifera* from the Argentine Andes I am obliged to J. Hunziker and O. Caso of Buenos Aires.

Several of the original figures and copied drawings were prepared by Mr. Stephen Tillett and Mrs. K. Goss.

Permission to use copyrighted illustrations was generously granted by four publishers. Some 18 of the figures in Chapter 4 were reproduced from *Illustrated Flora of the Pacific States*, vol. iii, by L. Abrams. For making these figures ɑvailable I am particularly indebted to Mrs. Roxana Ferris and the Stanford University Press. Figure 7 is reproduced from *Anatomy of the Dicotyledons* by C. R. Metcalfe and L. Chalk (Oxford University Press); Figure 23 was redrawn from *El Mundo Vegetal de los Andes Peruanos* by A. Weberbauer (Ministerio de Agricultura, Lima, Peru); and Figure 72 is taken from *The Genus Phlox* by E. T. Wherry (Morris Arboretum Monographs, Pennsylvania). The co-operation of the above mentioned publishers is gratefully acknowledged.

Many additional figures in Chapter 4 were taken from *Die Natürlichen Pflanzenfamilien* and *Das Pflanzenreich* (Verl. Wilhelm Engelmann, Leipzig). Figure 27 came from the University of California Publications in Botany, 1950. Individual figures in Chapter 2 were redrawn from a number of journals: American Journal of Botany, 1945, 1946; American Midland Naturalist, 1946; El Aliso, 1950, 1952; Proceedings of the National Institute of Sciences of India, 1940; and Rhodora, 1956. Figure 66 in Chapter 7 was first published in the journal Evolution.

Institutional support has been forthcoming from two sources. The Rancho Santa Ana Botanic Garden has supported the research on the Phlox family in every way possible throughout the years. A research grant from the National Science Foundation made possible much of the field and laboratory work incorporated in the new chromosome number records of Chapter 6, and helped in many other ways.

Claremont, California

January, 1957

CONTENTS

ILLUSTRATIONS

Frontispiece. Polemonium eximium on Mount Dana, California.

CHAPTER 6, CHROMOSOME NUMBERS

CHAPTER 7, PHYLOGENY

CHAPTER 8, PHYTOGEOGRAPHY

CHAPTER 9, GENERAL CONCLUSIONS

TABLES

CHAPTER I

SYSTEMATIC POSITION OF THE FAMILY

The historical task of blocking out the major taxonomic groups has proceeded in an uneven fashion. With regard to the Polemoniaceae the first minor taxon to be recognized was a species of *Polemonium* and the first of the larger taxa to achieve recognition were certain of the genera. The aggregation of related families, or order, became recognized at a later stage. Within these broad limits the family gradually emerged as a recognized taxon in the last years of the 18th century and the early years of the 19th.

The only known representative of the Phlox family was for some centuries the European species of *Polemonium, P. caeruleum.* In the 16th and 17th centuries this plant was usually called *Valeriana graeca* (viz., Dodonaeus 1559, Dewes 1578, C. Bauhin 1623, Gerarde 1633, Parkinson 1640, J. Bauhin and Cherler 1650). The name *Polemonium,* inherited by the herbalists from the Greeks, was associated with various other plants now going under the names *Silene, Lychnis, Gratiola, Valeriana, Trifolium* and *Papaver.* The true identity of the *Polemonium* of Dioscorides, though extensively discussed by the herbalists (see for example Parkinson 1640), has never been established with certainty, while Gerarde (1633) claimed that the *Polemonium* of Hippocrates was probably *Gratiola.* In 1700 Tournefort applied the name *Polemonium vulgare caeruleum* to the plant previously known as *Valeriana graeca,* and Tournefort's nomenclature was adopted by Linnaeus.

The process of describing the genera fell into four main periods. In the pre-Linnean period *Phlox* was set up by Plukenet (under the name of *Lychnidea*) in 1696, *Polemonium* by Tournefort in 1700, and *Loeselia* (as *Royenia*) by Houstoun about 1733.

Most of the genera of the Phlox family were established in the post-Linnean period from 1783 to 1839. During these years *Cantua* was described by Jussieu, *Cobaea* and *Bonplandia* by Cavanilles, *Gilia* and *Navarretia* by Ruiz and Pavon, *Ipomopsis* by Michaux, *Collomia* by Nuttall, *Linanthus* and *Eriastrum* (the latter as *Hugelia*) by Bentham, and *Leptodactylon* by Hooker and Arnott.

The third period of genus description was separated from the second by an interval of about half a century and lasted from 1896 to 1908. This period witnessed the addition of four minor genera, *Langloisia*, *Microsteris*, *Gymnosteris* and *Huthia*, by Greene and Brand. After an interval of another half century the minor genus *Allophyllum* was described in 1955 by Grant and Grant.

It is unnecessary to point out that the genera were not born full-fledged. In the case of a few of the genera, such as *Polemonium* and *Phlox*, the boundaries have been stable since the first, and subsequent history has consisted largely of the accumulation of knowledge about the species and their interrelationships. With regard to many of the genera, however, the working out of the generic limits has been a major task in itself. *Linanthus*, for example, was described in the early 19th century, but so were also *Leptosiphon*, *Fenzlia* and *Dactylophyllum*, which were destined to become united with *Linanthus* at a later date. *Ipomopsis* was described as a genus in 1803 but enjoyed only a shadowy nomenclatural existence for over a century and did not become clothed with the flesh of real species until 1936 and more especially 1956.

Linnaeus in the Genera Plantarum (1737) listed *Polemonium* close to *Convolvulus*, *Ipomoea*, *Datura* and *Campanula*; and put *Phlox* close to *Cordia*, *Azalea* and *Diapensia*. In Philosophia Botanica (1751) he placed *Polemonium* along with *Convolvulus*, *Campanula*, *Lobelia*, *Viola* and others in a group named the Campanacei, but relegated *Phlox* and *Loeselia* to a final list of "Vagae". Adanson produced a more natural, but still very rough, grouping in his Familles des Plantes (1763). *Polemonium*, *Phlox* and *Loeselia* were here united with *Bignonia*, *Convolvulus*, *Nicotiana* and numerous scrophulariaceous genera in a precursor of the modern Scrophulariaceae, the "Personées." It remained for Antoine de Jussieu in Genera Plantarum (1789) to first set the known polemoniaceous genera apart in a group of their own, the "Ordo Polemonia." In 1805 the family name was changed to its present form by Augustin Pyramus de Candolle in the Flore Française.

Jussieu's Polemoniaceae in 1789 consisted of *Phlox*, *Polemonium*, *Cantua* and *Loeselia* (as *Hoitzia*). Persoon enlarged it in 1805 by the inclusion of *Bonplandia*, *Gilia* and *Navarretia*, and Jussieu adopted a similar arrangement in 1810.

Cobaea had a varied history. This genus was first considered to be close to the Bignoniaceae by Cavanilles (1791). It was later transferred to the Polemoniaceae by Ventenat (1794). It was omitted from the Polemoniaceae, however, by Jussieu (1810), D. Don (1822) and Dumortier (1829). Persoon (1805) placed it in the Convolvulaceae, Jussieu (1810) in the Bignoniaceae, and D. Don (1824) removed it to a separate family, the Cobaeaceae. It was reintroduced into the Polemoniaceae by Lindley in 1830, removed again to the Cobaeaceae by G. Don (1838), and returned to the Polemoniaceae by Bentham (1845) and Lindley (1846), where it has remained ever since.

Three extraneous genera, *Diapensia* (Diapensiaceae), *Amphilophium* (Bignoniaceae) and *Cyananthus* (Campanulaceae) have had to be eliminated. The former was brought into the family by Ventenat in 1794 but was omitted from the treatments of Persoon (1805) and Jussieu (1810). *Amphilophium* was placed with *Cobaea* in the Cobaeaceae by Dumortier (1829) but this arrangement was not followed in the treatments of later authors. *Cyananthus* was introduced into the Polemoniaceae by Meisner in 1839 and Endlicher in 1840 and removed again by Bentham in 1845. These steps completed the task of building the foundations of our modern conception of the Phlox family.

Shortly after the establishment of the family as a separate taxon, efforts were begun to subdivide it into natural tribal units. These efforts have continued down to the present time. A conspectus of the evolving tribal and generic classification of the family is presented in Table 1.

Jussieu in 1789 placed the Phlox family next to the Convolvulaceae. This disposition, foreshadowed by Linnaeus' juxtaposition of *Convolvulus* and *Polemonium*, has been followed by nearly every subsequent author. Reichenbach (1837) went even further in subordinating the "Polemoniariae" along with the "Hydroleeae" as subfamilies of the Convolvulaceae. A more widely accepted arrangement proposed by Bartling (1830) was the classification of the Polemoniaceae together with the Convolvulaceae, Solanaceae, Hydrophyllaceae and Boraginaceae in a larger grouping, the Tubiflorae.

SYSTEMATIC RELATIONSHIPS

Relationships or at least resemblances suggestive of possibile affinities have been postulated between the Polemoniaceae and the Convolvulaceae[1], Hydrophyllaceae[2], Boraginaceae[3], Solanaceae[4], Geraniaceae[5], Caryophyllaceae[6], Primulaceae[7], Ericales[8], Bignoniaceae[9], Cucurbitaceae[10], Dipsacaceae[11], Fouquieriaceae[12], and Lennoaceae[13]. The characters upon which these suggestions have been based vary from general habit and more especially general floral

---

[1] Engler 1891; Alexnat 1922; Rendle 1925; Wettstein 1935; Erdtman 1952; Gundersen 1950; Soó 1953; Benson 1957.
[2] Engler 1891; Bessey 1915; Rendle 1925; Wettstein 1935; Hutchinson 1926; Gundersen 1950; Soó 1953; Benson 1957.
[3] Engler 1891; Bessey 1915; Wettstein 1935; Rendle 1925; Hutchinson 1926; Soó 1953.
[4] Engler 1891; Alexnat 1922; Hallier 1905; Bessey 1915; Wettstein 1935.
[5] Hutchinson 1926, 1948; Erdtman 1952; Wernham 1911–1912; Sundar Rao 1940.
[6] Dawson 1936; Wherry 1955: 67.
[7] Hallier 1905; Bessey 1915.
[8] Brown 1938.   [9] Flory 1937.   [10] Alexnat 1922.   [11] Alexnat 1922.
[12] Humboldt et al 1823; Engler and Gilg 1924; Flory 1937; Abrams 1951.
[13] Soó 1953.

Table 1. *Historical Development of a Tribal Classification of the Polemoniaceae*
(*Modern equivalents of the older generic names are given in parentheses*)

Reichenbach 1837	Meisner 1839	Bentham & Hooker 1876	Baillon 1890
Group Phloginae	Tribus Polemonieae	Group I	Tribe Polemonieae
1. Phlox	1. Bonplandia	1. Phlox	1. Polemonium
	2. Phlox	2. Collomia	2. Navarretia
Group Gilieae	3. Collomia	3. Gilia	3. Phlox
2. Welwitschia	4. Gilia	4. Polemonium	4. Collomia
(= Eriastrum)	5. Polemonium	5. Loeselia	5. Loeselia
3. Linanthus	6. Hoitzia (= Loeselia)	6. Bonplandia	6. Cantua
4. Leptosiphon	7. Cantua		
(= Linanthus)		Group II	Tribe Cobaeeae
5. Fentzlia (= Linanthus)	Tribus Cyanantheae	7. Cantua	7. Cobaea
6. Gilia	8. Cyananthus		
7. Aegochloa		Group III	Tribe Bonplandieae
(= Navarretia)	Tribus Cobaeaceae	8. Cobaea	8. Bonplandia
8. Courtoisia	9. Cobaea		
(= Collomia)			
9. Collomia			
10. Hoitzia (= Loeselia)			
11. Cantua			
Group Polemonieae			
12. Polemonium			
13. Cobaea			

Peter 1891	Brand 1907, 1908	Wherry 1940, 1945	Grant 1957
Tribe Cobaeeae	Subfam. Cobaeoideae	Polemonium Tribe	Tribe Cobaeeae
1. Cobaea	Tribe Cantueae	1. Polemonium	1. Cobaea
	1. Cantua	2. Polemoniella	
Tribe Cantueae	2. Huthia	3. Cobaea	Tribe Cantueae
2. Cantua			2. Cantua
	Tribe Cobaeeae	Collomia Tribe	3. Huthia
Tribe Polemonieae	3. Cobaea	4. Collomia	
3. Phlox		5. Bonplandia	Tribe Bonplandieae
4. Collomia	Subfam. Polemonioideae	6. Cantua	4. Bonplandia
5. Gilia	Tribe Polemonieae	7. Huthia	5. Loeselia
6. Polemonium	4. Polemonium	8. Navarretia	
7. Bonplandia	5. Collomia	9. Gymnosteris	Tribe Polemonieae
8. Loeselia	6. Phlox		6. Polemonium
	7. Gilia	Gilia Tribe	7. Allophyllum
	8. Aliciella (= Gilia)	Gilia Subtribe	8. Collomia
	9. Gymnosteris	10. Gilia	9. Gymnosteris
	10. Navarretia	11. Ipomopsis	10. Phlox
	11. Langloisia	12. Langloisia	11. Microsteris
		13. Eriastrum	
	Tribe Bonplandieae		Tribe Gilieae
	12. Loeselia	Linanthus Subtribe	12. Gilia
	13. Bonplandia	14. Leptodactylon	13. Ipomopsis
		15. Linanthastrum	14. Eriastrum
		16. Linanthus	15. Langloisia
			16. Navarretia
		Phlox Tribe	17. Leptodactylon
		17. Loeselia	18. Linanthus
		18. Phlox	
		19. Microsteris	

structure (Engler, Bessey, Hutchinson, etc.) to the more detailed features of nectar discs (Brown), chromosome size (Flory), floral anatomy (Dawson), pollen morphology (Erdtman), embryology (Sundar Rao), seed morphology, and serological reactions (Alexnat).

The origin of the family has been sought in the Rosales (Rendle 1925, Wettstein 1935); Caryophyllales (Bessey 1915, Dawson 1936); Geraniales (Wernham 1911–1912, Hutchinson 1926, 1948, Sundar Rao·1940); Malvales (Hallier 1905); Contortae (Soó 1953); Ericales (Brown 1938); Primulales (Hallier 1905, Bessey 1915); and Cucurbitales (Alexnat 1922). Clearly some critical appraisal is needed of these diverse points of view.

When we attempt to make such an appraisal, however, we come up against an insufficiency of morphological information. The relationships between modern plant families are greatly obscured due to our lack of knowledge of extinct groups. To a certain extent the gaps caused by extinction can be compensated for by detailed morphological studies of the living forms. For the Polemoniaceae and its relatives, however, such morphological surveys have not yet been carried to a point which will justify definitive phylogenetic conclusions. Outside of floral structure and general vegetative habit, our knowledge of the characteristics of the families concerned is still spotty and incomplete.

There are certain resemblances between the tropical woody Polemoniaceae and the Bignoniaceae. The winged seeds without endosperm and the minute rod-like chromosomes are similar in the two groups (Flory 1937). The woody vine-like habit, so common in the Bignoniaceae, is represented in the Polemoniaceae by *Cobaea*. Since in other respects the two families possess divergent specializations, which make unlikely the derivation of either one from the other, a possible explanation of their resemblances is that they are the markers of a common ancestry. But it is equally possible that the common characteristics in the seeds, chromosomes and growth habit are parallel developments in independent phylads.

The resemblances between the Polemoniaceae and the Fouquieriaceae in the corolla and gynoecium, again, may either be the result of real relationships or of convergences between quite distant orders. A comparative study of the Fouquieriaceae designed to clear up the systematic position of this anomalous family will be necessary to settle the question. In these and other cases of alleged family relationships a study and comparison of all the characteristics of the families concerned, not only floral structure but also wood anatomy, trichomes, embryo and seedling structure, chromosome number and morphology, and the like, would provide a sound basis for decisions concerning phylogenetic affinities[14].

[14] A model study of the systematic relationships of a family, making use of all characters, is Moseley and Beeks' (1955) work on the Garryaceae and their relatives.

For the present, much can be said for holding to the time-honored practice of placing the Polemoniaceae between the Convolvulaceae and Hydrophyllaceae. The inflorescence (Rendle 1925), corolla, stamens, nectaries (Brown 1938), and orientation of the ovules (Engler 1891) are much the same in the Convolvulaceae and Polemoniaceae. The aestivation of the corolla in bud is of a similar spirally twisted and infolded type in these two families. The similarities in pollen morphology have been pointed out by Erdtman (1952). The convolvulaceous genus *Pharbitis* has a 3-carpellary gynoecium like the Polemoniaceae, and the polemoniaceous genus *Cobaea* approaches the Convolvulaceae in its vine-like habit and calyx split to the base. The relationship between the Polemoniaceae and Hydrophyllaceae is attested to by many similarities of the vegetative habit and general floral plan.

The identification of the closest modern equivalent of the ancestral group from which the Polemoniaceae has been derived is an even more difficult problem. It is a problem which in the absence of a fossil record can only be discussed in hypothetical terms.

The view held by Bessey that the group is derived from the Caryophyllales seems improbable. Brown (1938) has pointed out the futility of attempting to derive the woody tropical members of the Tubiflorae from the temperate herbaceous Caryophyllales. Furthermore, the central or basal placentation and the position of the nectaries at the base of the stamen whorls in the Caryophyllales do not qualify them as likely ancestors of groups with predominantly axile placentation and a nectary disc attached to the base of the ovary (Brown 1938). Similar arguments exclude the Primulales, which seem to be more specialized derivatives of the Caryophyllales (Dickson 1936, Douglas 1936), as ancestral forms for the Tubiflorae.

Hutchinson's suggestion that the Polemoniaceae are derived from the Geraniales would likewise be very difficult to defend. The temperate distribution, herbaceous habit and staminal nectaries of the Geraniales practically disqualify them as an ancestral plexus from which the tropical woody Tubiflorae with their very different type of nectary might have emerged. Where the Geraniales do fit into the phylogenetic system of the Angiosperms is not altogether clear. I am inclined at present to accept Bessey's view that they are a side branch from a Ranalian stock which has not served as the starting point of any extensive new evolutionary developments.

Wettstein's idea of a common origin of the Tubiflorae and Contortae is worthy of serious consideration. Absolute characters for distinguishing the two orders are difficult to find (Benson 1957). Within the Contortae the tropical woody family Loganiaceae possesses more primitive characteristics in general than the temperate herbaceous Gentianaceae and Asclepiadaceae. Any phylogenetic connection of the Tubiflorae with the Contortae should be most apparent in comparisons of the primitive representatives of the two orders.

Brown (1938) has noted that the general characters of the Ericales and Tubiflorae are very similar and that the woody members of the latter order are but "an easy step" from the more generalized Ericales. As an example of a primitive and generalized member of the Ericales *Clethra* is cited, though Brown cautions that "We do not necessarily have to assume that the Clethraceae and the Ericaceae are the actual transitional forms. In fact, the peculiar stamens of this group argue against this assumption". There is no great difference, furthermore, between the Clethraceae and the Dilleniaceae or Theaceae which can not be accounted for by the known trends towards reduction and fusion of parts (Lechner 1914, Brown 1938).

Whether the Clethraceae in particular or the Ericales in general can be regarded as a good choice of a transitional condition is open to question. The poricidal anthers, the vascular anatomy of the flower (Kavaljian 1952), and the type of inflorescence in *Clethra* are specialized features which isolate the group from the primitive Tubiflorae. Some part of this difficulty may be avoided, as Dr. G. L. Stebbins has pointed out to me, by looking for the primitive relatives of the Tubiflorae in the Styracaceae or elsewhere in the Ebenales rather than in the Ericales.

The tentative picture which emerges is, in any case, an evolution of woody plants in a tropical setting from a Ranalian ancestor through successive stages of floral and vegetative development marked by the characters of the present Theales and Ebenales (or Ericales) to the common beginnings of the Contortae and Tubiflorae. The subsequent diversification of the tubiflorous families, the Convolvulaceae, Polemoniaceae, Hydrophyllaceae, Boraginaceae, Solanaceae, etc., was followed then by independent development of herbaceous forms in each family.

CHAPTER II

MORPHOLOGY

The methods and objectives of the phylogenetic taxonomist and the comparative morphologist differ only in the depth of the analysis. In his actual study of characters, and excluding for the moment his other occupations such as field work and nomenclature, the taxonomist is a pioneering comparative morphologist. He explores a large territory, more or less superficially by later standards, and opens it up for the more thorough and intensive activities of the comparative morphologists who follow him.

The purpose of either the phylogenetic taxonomist or comparative morphologist is to discover the natural relationships of higher categories. The result of his work in either case is a system of classification which may or may not be the most convenient one for identification and pigeon-holing but which does reflect the best available evidence of evolutionary affinities. The basic assumption is that some, but not all, morphological resemblances between distinct organic groups are unlikely to be the result of chance or convergent evolution, and hence reflect phylogenetic relationships; and that some, but not all, differences are due to evolutionary divergences from common ancestral stocks and hence mask phylogenetic relationships.

The morphological characters known to be of taxonomic value in the Polemoniaceae are summarized in Chapter 4. These characters, incorporated into the tribal, generic and sectional diagnoses, are the cumulative product of several generations of taxonomic work, with some desultory assistance from the field of morphology. They reveal the main lines of relationship but lack analytical exactness. The Phlox family is in a state of exploration now which is ripe for comparative morphological work. There is scarcely a character – internal morphology of the seed, structure of the seed coat, cotyledons, corolla venation, pollen grains, etc. – which would not repay close comparative study.

It is hoped that one of the functions of the present taxonomic review will be to provide the framework and stimulus for such studies. To this same end a living collection of Polemoniaceae is being assembled at the Rancho Santa Ana Botanic Garden which is available for morphological investigations. It is my pleasant duty, in this connection, to thank the various colleagues in Mexico, South America and Europe who have sent seeds and thus supplemented our predominantly western North American collections in a most important way.

EMBRYO AND SEEDLING

The embryo lies in the middle of the endosperm (Brand 1907, Martin 1946) and has a spatulate form in the species so far investigated (Fig. 1). Brand found that in *Cobaea* and *Cantua* the embryo is colorless with a very small radicle and large cordate or ovate cotyledons swollen with nutritive tissue. In the temperate herbaceous genera, by contrast, he found the embryo to be green or greenish (or sometimes colorless), mostly with a radicle equalling or exceeding the cotyledons in size, and with foliaceous, linear or obtuse cotyledons lacking stored food materials.

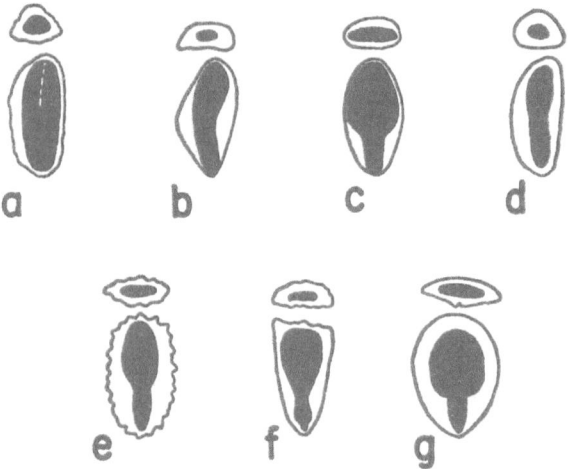

Fig. 1. Form and position of the embryo in the seed.
a) Navarretia squarrosa. *b*) Ipomopsis aggregata. *c*) Microsteris gracilis. *d*) Polemonium foliosissimum. *e*) Phlox pilosa. *f*) Linanthus ciliatus. *g*) Collomia grandiflora. (from Martin.)

The embryogeny of *Phlox drummondii, Polemonium caeruleum, Polemonium pauciflorum,* and *Gilia tricolor* has been followed in detail by Miller and Wetmore (1945a) and Souèges (1939a, 1942, 1945). The general pattern of development is the same in all examples studied.

In *Phlox drummondii* fertilization usually occurs 12 to 36 hours after polli-

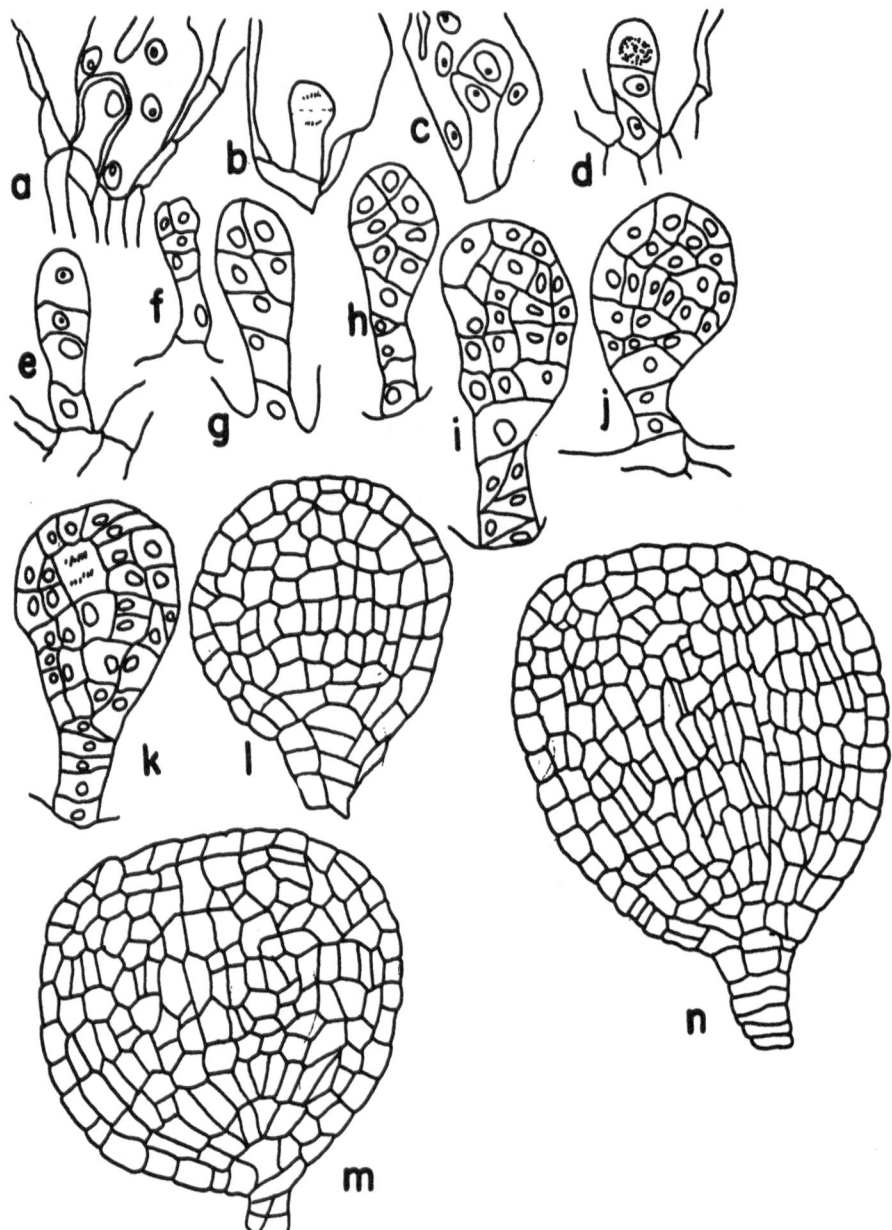

Fig. 2. Early development of the embryo in Phlox drummondii.
a) Zygote surrounded by free nucleate endosperm. *b*) Zygote at first division, 22 hours. *c*) Two-celled filamentous proembryo, 1 day. *d*) Three-celled filamentous proembryo, 1 + days. *e*) Four-celled filamentous proembryo, 1 + days. *f–i*) Proembryo, 1–3 days. *j–k*) Proembryo, 4th day. *l*) Proembryo, 5th day. *m*) Proembryo, 6th day. *n*) Proembryo, 8th day. (from Miller and Wetmore; × 300.)

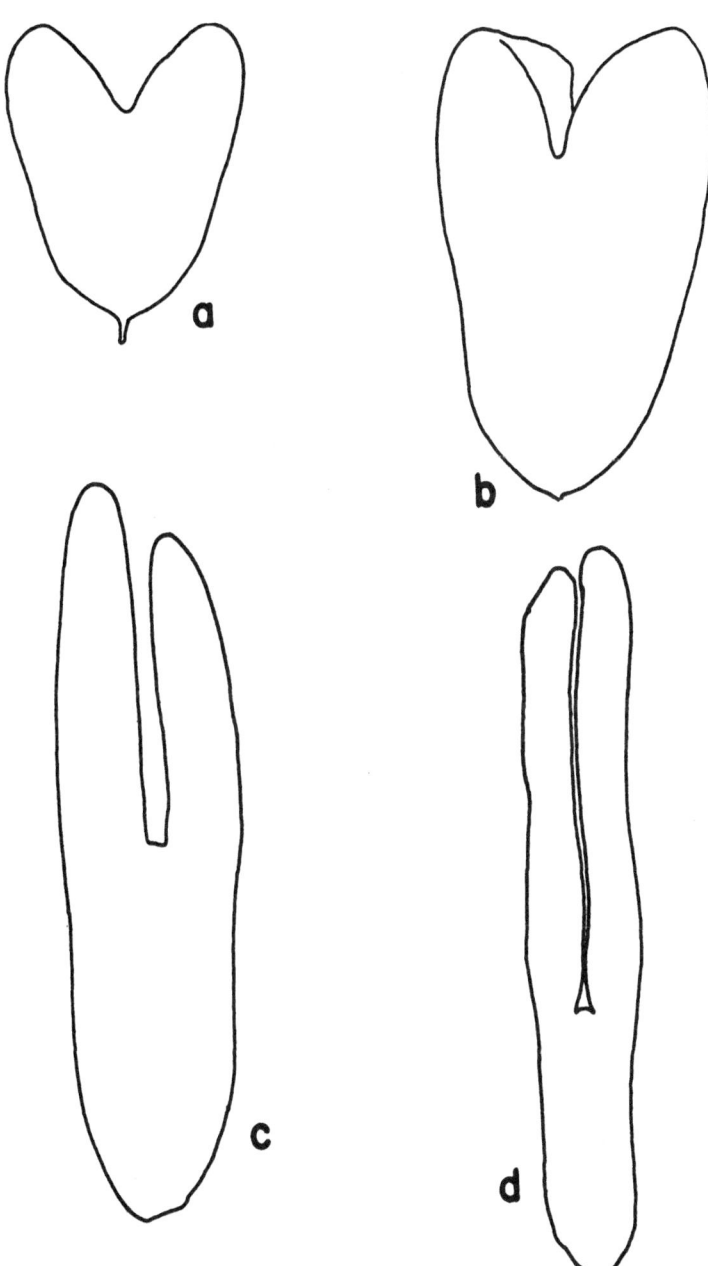

Fig. 3. Development of the embryo in its older stages in Phlox drummondii. Cellular detail omitted. *a*) Heart-shaped embryo, 8th day. *b*) Torpedo-shaped embryo, 9th day. *c*) Torpedo-shaped embryo, ± 9th day. *d*) Mature embryo, 30 days. (from Miller and Wetmore; *a–c* × *83; d* × 14.)

nation. Following fertilization the endosperm nucleus divides actively giving rise to a free-nucleate endosperm which later becomes cellular by centripetal wall formation (Fig. 2a). The zygote begins to divide at the time of wall formation in the endosperm about 22 hours after fertilization.

The embryo passes through a short-lived filamentous stage on the first day (Fig. 2b–e), a proembryo stage when 1–8 days old (Fig. 2f–n), a heart-shaped stage on the seventh to ninth days (Fig. 3a), a torpedo stage from the ninth to the fifteenth day (Fig. 3b–c), and becomes a mature embryo when 30 days old (Fig. 3d).

The first three divisions of the zygote are in the same plane and give rise to a linear tier of four cells which constitutes the filamentous stage of the embryo. The apical cell of the filament next divides into two longitudinal halves, initiating the proembryo stage. The proembryo eventually becomes a mass of cells above a stalk-like suspensor. As a result of the development of primordia of the cotyledons, the embryo next assumes a heart-shaped form. Elongation of the cotyledons and appearance of a definite hypocotyl mark the so-called torpedo stage. A month after fertilization the parts of the embryo have reached the dimensions which they will retain in the mature seed.

A differentiation of the tissues goes hand in hand with the enlargement of the embryo and development of its major parts. For a description of the anatomical and histological details the reader is referred to the original paper of Miller and Wetmore.

Twin embryos were found in a seed of *Phlox speciosa* by Brand (1907: 13).

Miller and Wetmore (1945b) showed that in *Phlox drummondii* germination is preceded by cell elongation in the lower part of the mature embryo. After germination the seedling grows by meristematic activity in both the root tip and shoot tip. The root at first elongates more rapidly than the shoot. A core

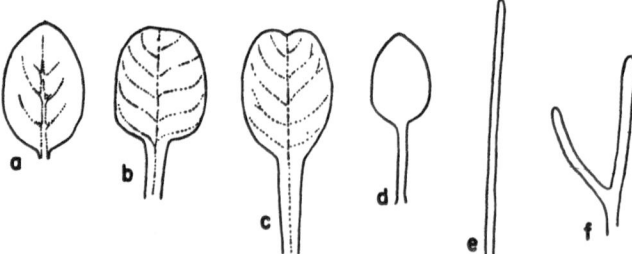

Fig. 4. Cotyledons.
a) Cantua pyrifolia. b) Collomia grandiflora. c) Allophyllum divaricatum. d) Gilia achilleaefolia multicaulis. e) Gilia ochroleuca bizonata. f) Navarretia pubescens. (a–e original, by A. Grant; f redrawn from Brand; a–e × 2; f × 4.5.)

of procambium, present already in the young embryo, advances toward the apices of the growing root and shoot.

The cotyledons are with but few exceptions entire-margined. In *Collomia* and *Allophyllum* they are broadly ovate, somewhat leathery in texture, marked by prominent raised veins, and frequently possess an apical notch (Fig. 4b–c). The cotyledons of *Cantua* have a similar shape and venation (Fig. 4a). In *Gilia* they vary from ovate in the distal region to narrowly linear, and are composed of soft tender tissues with faint veins (Fig. 4d–e). An exceptional condition in *Navarretia pubescens*, *N. setiloba* and *N. mitracarpa* is the presence of three or four cotyledons which are divided to the base into linear segments (Brand 1907: 2, 13; see Fig. 4f).

THE VEGETATIVE BODY

The vegetative body develops, as Miller and Wetmore (1946) have shown for *Phlox drummondii*, by the activity of growing points which have descended in an unbroken lineage from the meristems present in the young embryo. The root as we have already seen undergoes a rapid growth in the first stages after germination. Shortly thereafter the meristem in the region between the cotyledons grows upward and commences to lay down foliar primordia. The apical growing points of both the root and the shoot can be traced back to meristems of the lower and upper hypocotyle in the young embryo.

The shoot apex of *Phlox drummondii* consists of a corpus or central mass of cells with walls oriented in various planes covered by a tunica or mantle of regularly arranged cells one or two layers thick. The shoot apex gives rise to leaf primordia in decussate pairs. The foliar appendages are then extended by cell division and cell enlargement, each leaf primordium elongating by meristematic activity at its apex. Vegetative buds commence to form in axillary points three or four nodes below the shoot apex.

A procambial strand develops acropetally into the foliar primordium from the continuous vascular system below. About the time that the procambium of a leaf becomes evident a leaf gap begins to appear on the leaf directly below. The procambium running into a bud develops acropetally from the lateral margins of the leaf and stem vascular tissue at the node. The node of a mature shoot, showing leaf traces and leaf gap, is illustrated in Figure 5.

The growth of the stem is continuous from the seedling stage on in *Cobaea* and *Cantua*, but is interrupted by a period of root and basal leaf development in many temperate herbaceous genera. In *Polemonium* this interruption lasts for a few weeks in some species (*P. flavum*, *P. pauciflorum*) or for nearly a year in others (*P. reptans*, *P. caeruleum*). In the former case the plant goes on to flower in the first year, while in the latter case the stem development and flowering occur in the second year. The biennial and perennial species of *Ipomopsis* produce a basal rosette of leaves in their first year of growth which is followed by stem development and flowering in the second season. Among plants with an annual life cycle in *Gilia* some species have continuous stem growth and others a temporary rosette stage.

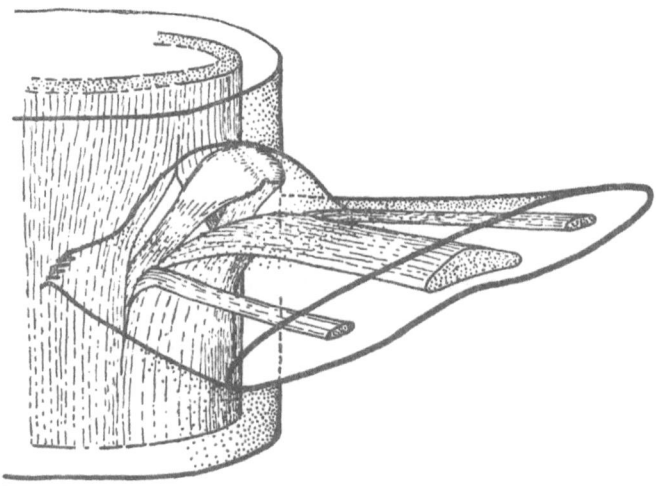

Fig. 5. Node in Phlox drummondii.
The main and two lateral leaf traces are diverging from the vascular tissue of the stem. A young vegetative bud is above the leaf traces. The leaf gap is visible in the vascular cylinder behind the bud. (from Miller and Wetmore; × 40.)

The mature plant may be a small tree (*Cantua*), woody vine (*Cobaea*), shrub (*Cantua*), subshrub (*Leptodactylon, Phlox*), perennial herb (*Polemonium, Phlox, Ipomopsis*), or annual herb (*Gilia, Linanthus, Navarretia*, etc.). The plant body ranges in size from lianes 24 feet or more high in *Cobaea*, and trees up to 12 feet tall in *Cantua*, to delicate leafless annuals one or two centimeters tall in *Gymnosteris*. The great majority of the Polemoniaceae are herbaceous plants of medium size.

The trichomes on the stems and leaves are of several distinct types. Multicellular gland-tipped hairs and multicellular villous hairs occur widely throughout the family. Each type has many modifications. The stipitate glandular hairs are mostly short and stout (Fig. 6a), but in *Polemonium reptans villosum* are exceedingly long and slender (Fig. 6b). The cells composing the villous hairs are normally cylindrical in form and the trichome itself is a long tapering cone (Fig. 6c). In *Gilia capitata* the individual cells are sometimes rectangular parallelepipeds instead of cylinders and are joined at right angles to one another like the links of a chain (Fig. 6d). The glandular hairs of *Polemonium, Allophyllum, Collomia, Gilia,* and *Navarretia* are frequently associated with the elaboration of a skunk-like odor.

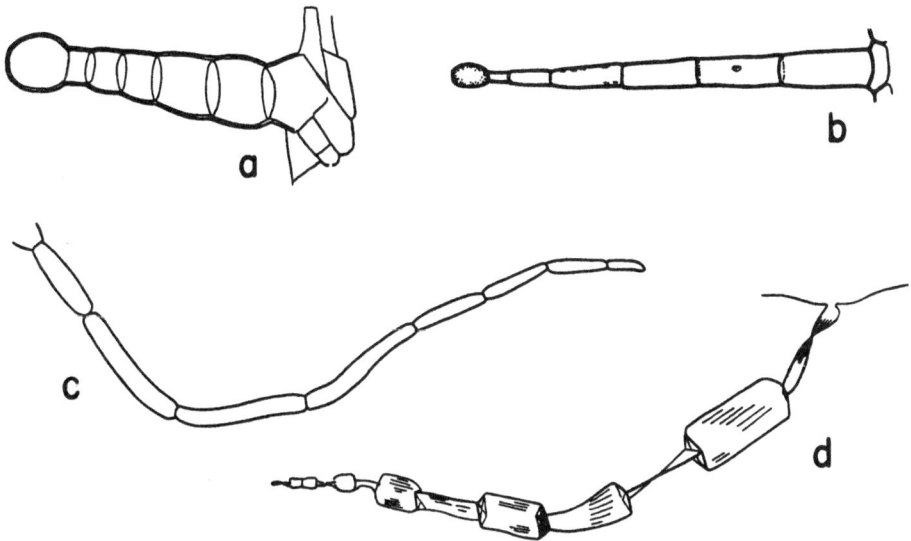

Fig. 6. Trichomes.
a) Short stout glandular hair (Gilia capitata). *b)* Long slender glandular hair (Polemonium reptans villosum). *c)* Villous hair (Gilia capitata). *d)* Chain-like hair (Gilia capitata). (*a, c, d,* original; *b* redrawn from Braun; actual length: *a,* 0.12 mm.; *b, c,* ca. 1 mm.; *d,* 2.4 mm.)

The distribution of different types of trichomes is often found to be of systematic value. The abundance and length of the glandular hairs serve to distinguish *Polemonium reptans villosum* from *P. r. reptans. Gilia stellata* is distinguished from related species and indeed from the rest of the genus *Gilia* by the presence of geniculate hairs. A very fine arachnoid pubescence helps to define the section *Arachnion* of *Gilia.* The genera *Eriastrum* and *Navarretia* differ among other things by the fact that the former is densely wooly but never glandular while the latter is frequently very glandular but is never densely wooly. The foregoing incomplete remarks suggest that much could be learned by a detailed study of trichomes in the Polemoniaceae and extensive survey of their systematic distribution, as have been carried out in other families of Angiosperms.

The following notes regarding the stem have been set forth by Metcalfe and Chalk (1950: 941–942; see also Fig. 7). The cortex of the primary system of stem tissues is often partly collenchymatous, sometimes includes mucilaginous cells, and contains stone cells in *Phlox paniculata.* The cork is believed to arise in the outer part of the cortex or in the pericyclic region in *Phlox.* The endodermis is well defined in at least some of the Polemoniaceae. It

consists of large, tangentially elongated cells in *Collomia linearis*, uniformly
thickened cells in *Eriastrum densifolium*, and of cells provided with definite
Casparian thickenings in *Phlox paniculata* and *Polemonium caeruleum*. Inconspicuous pericyclic fibers were found in some species. The phloem is present
either as a continuous ring without mechanical elements or as separate strands
embedded in sclerenchymatous tissue and in contact with the xylem. Solereder
saw a yellow secretion in the intercellular spaces of the phloem in *Cobaea
scandens*. The pith may contain sclerenchymatous cells like those in the
cortex.

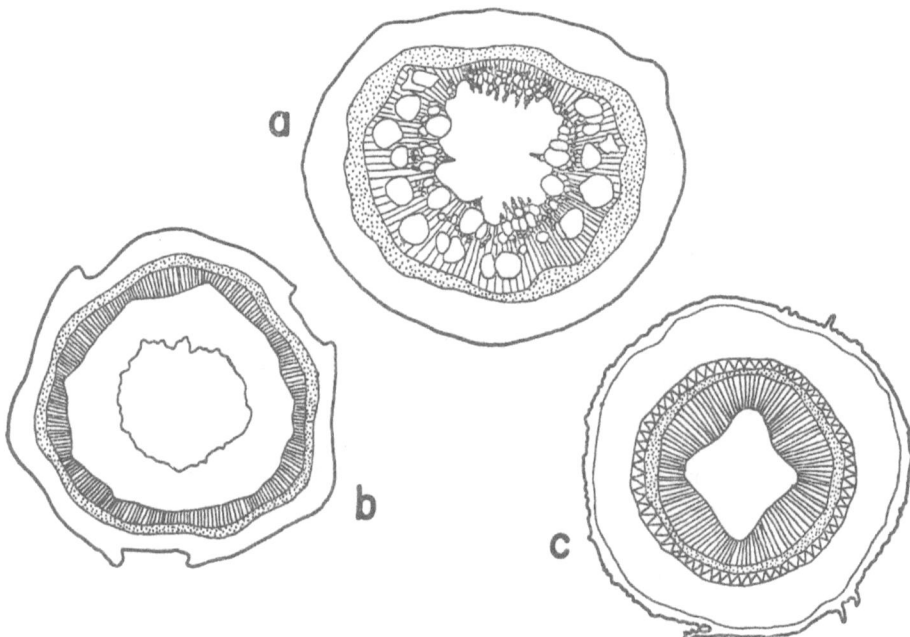

Fig. 7. Stem in transverse section.
The regions proceeding from the outside to the center are as follows: cortex, shown white; cork
(in *c* only) shown by zigzag lines; phloem, stippled; xylem, radial lines; and pith, white. The
xylem is shown with large vessels in *a*. The central part of the pith has disappeared in *b*. There is
an inner and outer part of the cortex in *c*, the outer cortex being composed of small parenchymatous cells and the inner of large hard-walled cells.
a) Cobaea scandens. *b*) Polemonium caeruleum. *c*) Phlox subulata. (from Metcalfe and Chalk;
magnified.)

The xylem in all of the species examined by Metcalfe and Chalk is present
in a continuous cylinder without well-defined medullary rays. The vessels
are in radial rows. Wood fibers with simple or inconspicuously bordered pits
and uniseriate rays were found in *Collomia* and *Loeselia mexicana*.

In most species so far examined the vessel elements are rather small, being
about as wide as the surrounding parenchymatous cells. Solereder recorded

a diameter of about 40 microns for the vessel elements of some woody species and Crampton (1954) gave the diameter as 20 microns in the small annuar *Navarretia propinqua*. Large vessels 200 and more microns in diameter occul

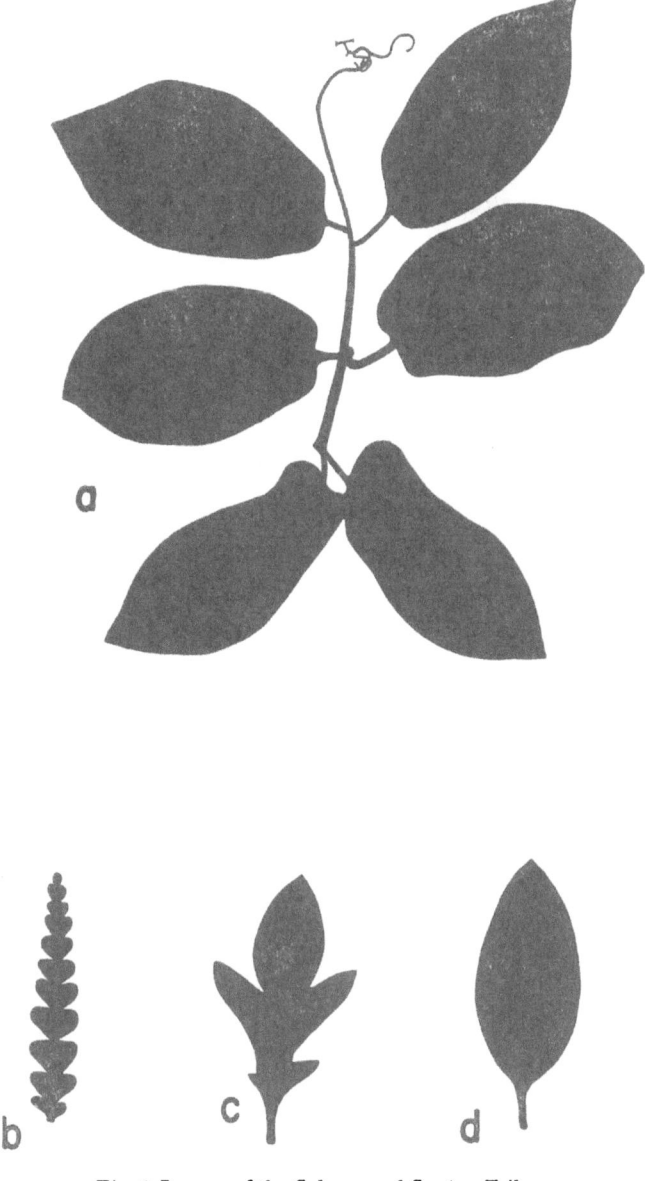

Fig. 8. Leaves of the Cobaea and Cantua Tribes.
a) Cobaea scandens. *b*) Huthia coerulea. *c*) Cantua buxifolia. *d*) Cantua pyrifolia. (original, by S. Tillett; natural size.)

in *Cobaea scandens*. Crampton (1954) measured the average length of vessel elements in several species of *Navarretia* with the following results: 0.8 mm. in *N. propinqua*, 0.7 in *N. involucrata*, 0.4–0.5 mm. in the majority of species, and 0.25 in *N. tagetina*.

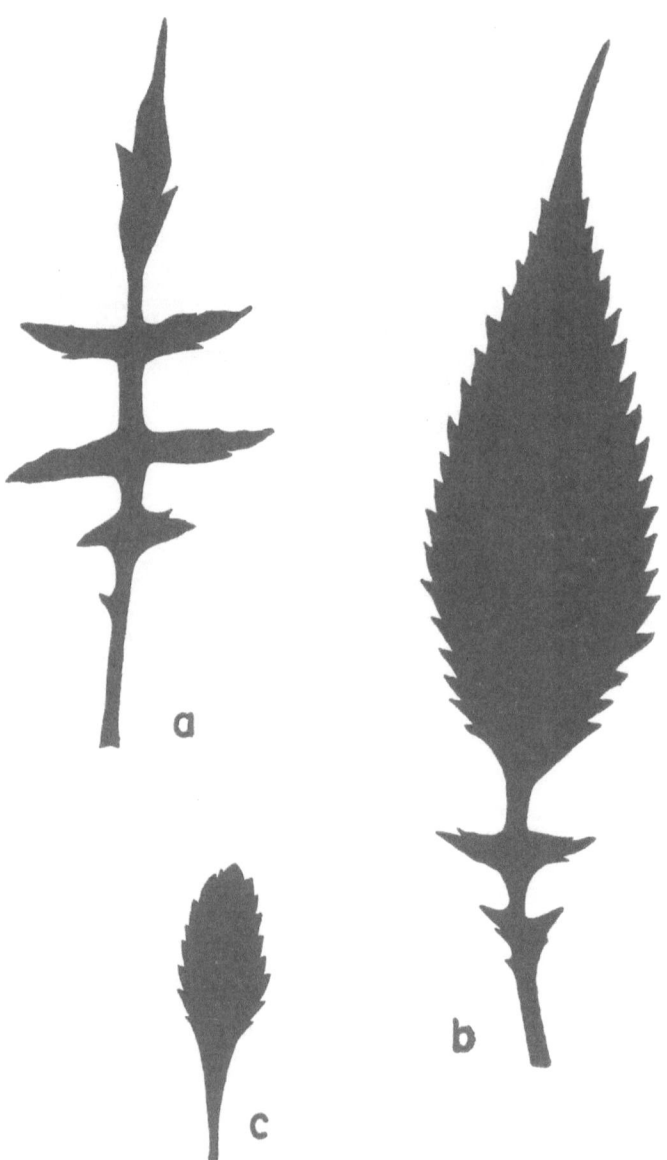

Fig. 9. Leaves of the Bonplandia Tribe.
a) Bonplandia linearis. *b*) Bonplandia geminiflora. *c*) Loeselia mexicana. (original, by S. Tillett; natural size.)

Particular interest is attached to the perforations of the vessels because of the well established phylogenetic sequence in regard to this feature. It appears that a wide range of conditions from primitive to advanced is repre-

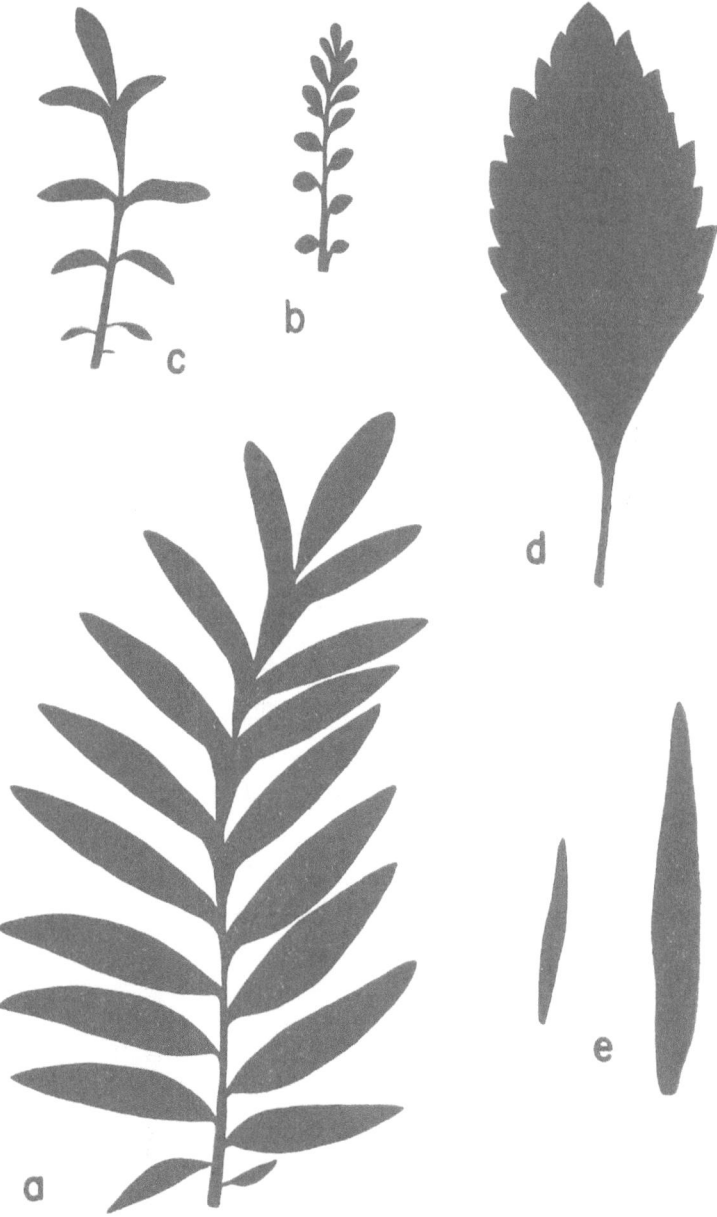

Fig. 10. Leaves of the Polemonium Tribe.
a) Polemonium foliosissimum. *b*) Polemonium micranthum. *c*) Allophyllum divaricatum. *d*) Collomia rawsoniana. *e*) Collomia linearis. (original, by S. Tillett; natural size.)

sented within the bounds of the Phlox family. As yet, however, no comprehensive study of this aspect has been undertaken. Here is another fruitful
subject for future investigation. The few desultory facts so far gathered can

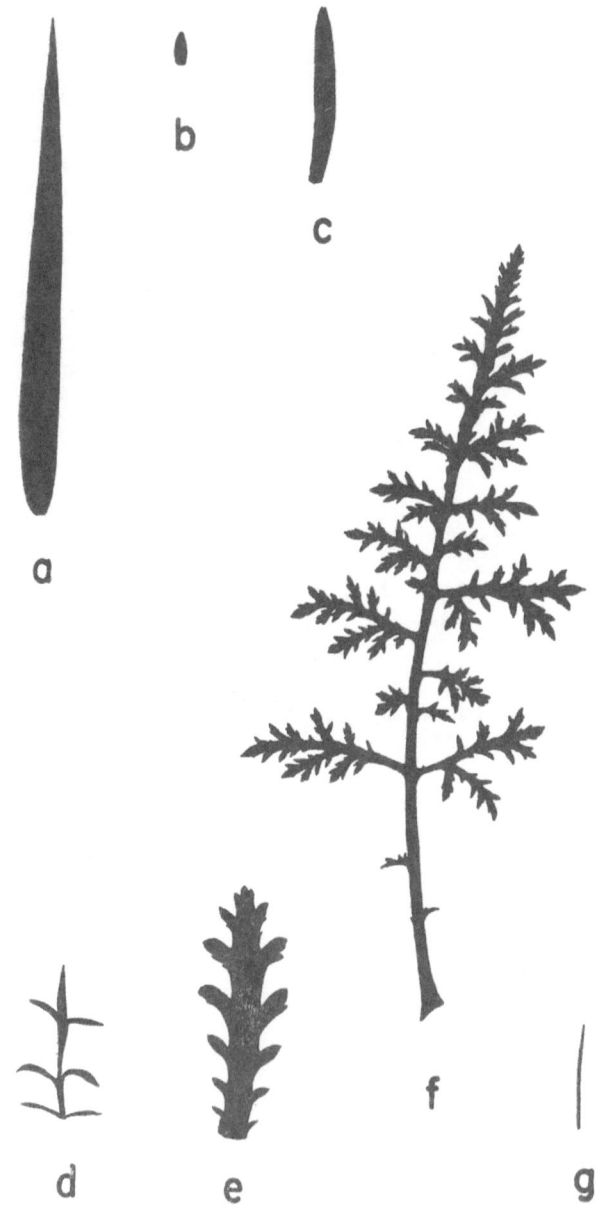

Fig. 11. Leaves of the Polemonium Tribe (cont.) and Gilia Tribe.
a) Phlox glaberrima. *b*) Phlox covillei. *c*) Microsteris gracilis. *d* Gilia rigidula. *e*) Gilia subnuda.
f) Gilia splendens. *g*) Gilia leptalea. (original, by S. Tillett; natural size.)

be briefly summarized. Both reticulate and scalariform perforations with two
to six bars, together with porous perforations, occur in *Navarretia squarrosa*

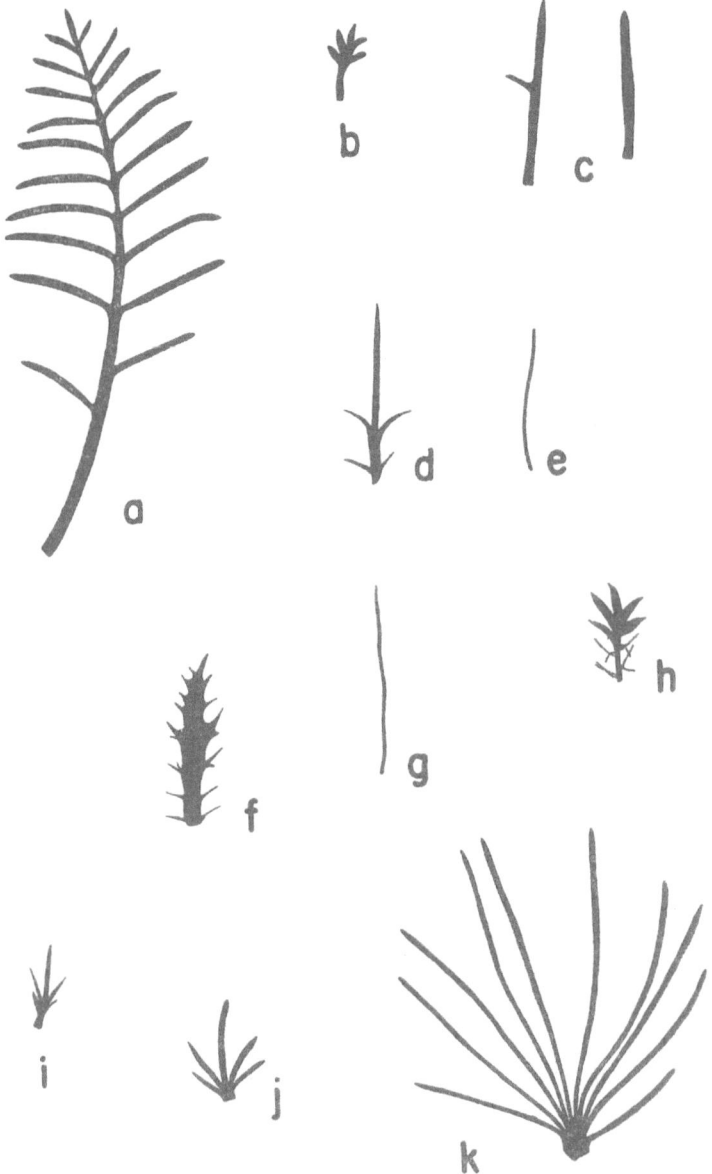

Fig. 12. Leaves of the Gilia Tribe (cont.).
a) Ipomopsis aggregata. *b*) Ipomopsis congesta. *c*) Ipomopsis tenuifolia. *d*) Eriastrum densifolium.
e) Eriastrum filifolium. *f*) Navarretia atractyloides. *g*) Navarretia minima. *h*) Langloisia punctata.
i) Leptodactylon pungens. *j*) Linanthus nuttalli. *k*) Linanthus grandiflorus. (original, by S.
Tillett; natural size.)

and *N. atractyloides*; in the other species of *Navarretia* only porous perforation plates, a derived condition, are found (Crampton 1954). Vessels with simple perforations together with some having scalariform plates with one to ten bars were found in the xylem of *Loeselia* and *Polemonium* by Solereder. The vessel perforations are mostly simple in *Cobaea, Loeselia, Collomia* and *Phlox* (Metcalfe and Chalk 1950).

The root in *Phlox ovata* has a hypodermis consisting of radially elongated, pentagonal cells, a cortex composed of 12–20 rows of parenchymatous cells some of which contain starch grains, and a triarch to pentarch vascular structure (Kraemer 1910).

The leaves are alternately arranged in the majority of the Polemoniaceae. In *Phlox* and *Linanthus* the leaves are opposite, but even here the uppermost leaves and the bracts are sometimes alternate or subopposite. In most of the genera, again, the upper stem leaves are not differentiated from the lower stem leaves in regard to the degree and manner of development. In *Gilia*, however, the upper cauline leaves are more or less strongly reduced, giving the whole plant a scapose habit, whereas in *Eriastrum, Langloisia* and *Navarretia* the upper stems are frequently leafy and the lower stems naked. *Gymnosteris*, as we have already seen, has persistent cotyledons and involucral bracts but no true foliage leaves on the mature shoot.

In form the leaves are pinnately compound in *Cobaea* and *Polemonium*, simple with a pinnate venation in all but two of the other genera, and palmately divided in *Leptodactylon* and *Linanthus*. The leaves where simple are usually serrate, dentate or pinnately lobed or cleft in various ways; entire-margined leaves with a broad ovate blade are found more rarely, as in *Cantua pyrifolia*. A common trend of reduction within the herbaceous genera ends with the production of narrow linear or even needle-like leaves. An overview of the principal leaf forms in the family is presented in Figures 8–12.

Metcalfe and Chalk (1950: 941) characterize the cuticle of the leaf as "smooth, finely striated, granular or verrucose." The epidermis of narrow or linear leaves, according to these authors, is composed of cells elongated in the direction of the median vein. The outer walls of the leaf epidermis tend to be mucilaginous in species of *Collomia, Phlox* and *Gilia*, and isolated marginal or apical cells of the epidermis present papillae, sometimes accompanied by silicified bodies, in some species of these same genera and in *Bonplandia* and *Loeselia*. The stomata are "generally ranunculaceous, but [tend] to be rubiaceous in species with narrow leaves." They are usually present on both leaf surfaces, but rarely are confined to the lower side. *Phlox hoodii* with reduced needle-like leaves has stomata developed only in the basal region where they are protected by hairs. The mesophyll of the leaves is usually devoid of large intercellular spaces. Large aqueous cells are found around the

vascular bundles. The veins have parenchymatous sheaths, which are sometimes suberized, and sclerenchymatous cells, along with the vascular cells.

INFLORESCENCE AND FLOWER

The apical meristem destined to give rise to flowers has been compared
with the vegetative apex in *Phlox drummondii* by Miller and Wetmore (1946).
They find that the two types of apex are generally similar in the presence of
a tunica-corpus organization of tissues and in the acropetal development of
procambium. The floral apex, however, is more rounded, has three or four
rows of cells in the tunica instead of the one or two rows characteristic of a
vegetative apex, and has a relatively smaller corpus region. A further difference
lies in the arrangement of the foliar appendages. A change from a decussate
phyllotaxis to a nearly cyclic arrangement of parts marks the change from a
vegetative to a floral apex.

The floral bud also develops faster than a vegetative bud and in the course
of this development its meristematic tissues are used up in the formation of
floral appendages. The sepal primordia develop first on the apex of the floral
bud, followed in turn by the primordia of the stamens, carpels and petals.
The procambium, which has extended into the floral bud forming a cylinder,
now extends acropetally into each of the individual floral parts.

In most members of the Phlox family the inflorescence is of the determinate
type in which a terminal flower blooms first and the other flowers derived
from axillary buds below the terminal one bloom subsequently in a centrifugal
order. The number of flowers in each cymose cluster subtended by a bract
varies from two to several or more rarely to many. The larger inflorescences
usually comprise aggregations of numerous cyme-like units borne on one or
more main stems, and hence are compound to varying degrees. A very detailed
account of this general sequence of flowering in *Phlox paniculata* has been
given by Wagner (1901). The flowers are axillary in the vegetative shoot in
Cobaea, Huthia, Bonplandia, and *Loeselia.*

Inflorescences consisting of solitary flowers are found in *Cobaea* and *Phlox*
sect. *Occidentales*. The ecological effectiveness of the inflorescence is very
different, however, in the two cases. In *Cobaea* the individual flowers are
stretched out in the leaf axils along the length of the stem and are separated
from one another by long internodes, whereas in the cespitose Phloxes the
solitary terminal flowers of many different vegetative shoots are aggregated
in a dense mass of blooms raised just above the cushion-like plant body.
At the opposite extreme many-flowered compound inflorescences consisting
of dense glomerules, heads or capitula are found in *Collomia, Gilia, Ipomopsis*,
and other genera.

Floral bracts are lacking in the tropical genera and in some temperate genera but are developed in *Collomia, Gymnosteris, Phlox, Ipomopsis,Eriastrum, Navarretia* and *Langloisia*. Brand observed that a correlation exists between the presence of peduncled flowers and the absence of bracts.

The flowers are hermaphroditic, though male-sterile flowers are formed on plants belonging to several herbaceous genera under conditions of extreme stress or difficulty.

The flowers are normally pentamerous with a 3-carpellary ovary and 3-lobed stigma, but here again exceptions exist. In the derivatives of an interspecific hybrid in *Gilia*, 3, 4, 6 and 7-merous flowers were observed (Grant 1956). A garden strain of *Phlox subulata* of uncertain parentage likewise showed frequent deviations from the normal 5-merous condition, 12% of 8389 flowers exhibiting 4, 6 or 7 corolla lobes (Vuillemin 1909). Similar conditions are met with as rare mutants in the natural populations of several species; and a deviating floral plan has become fixed as a new norm in a few populations. *Navarretia cotulaefolia, Leptodactylon caespitosum* and some populations of *Phlox sibirica* have tetramerous flowers and *Leptodactylon jaegeri* has hexamerous flowers. Two-cleft stigmas occur in several species of *Navarretia* including *N. cotulaefolia*.

The calyx is divided nearly to the base in *Cobaea*, but is definitely fused into a tube in the other genera. It is herbaceous in all the tropical genera except *Loeselia* and in *Polemonium* and *Collomia* among the temperate genera. In the remaining genera the green tissue is confined to the calyx lobes while a membranous, often hyaline, tissue is developed in the sinuses between adjacent lobes (Fig. 13). The presence or absence of a differentiation of tissues in the calyx, as well as the specific form of the sinuses and lobes, have since the time of Greene been usefully employed as systematic characters.

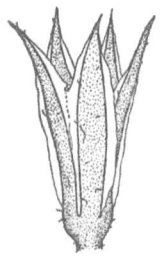

Fig. 13. Calyx of Gilia tricolor. (from Grant; × 5.)

The corolla is usually radially symmetrical but is not uncommonly slightly bilabiate and in a few cases, such as *Langloisia matthewsii*, is quite strongly bilabiate. The form of the corolla is rotate in *Polemonium*, campanulate in *Cobaea* and *Polemonium* (other species), tubular in *Cantua* and *Ipomopsis*, funnelform in *Gilia* and *Navarretia*, and salverform in *Phlox* and *Linanthus*. An impression of the diversity of corolla types in the family can best be obtained through a perusal of the systematic treatment and the accompanying illustrations in Chapter 4.

The stamens are epipetalous throughout the family. The point of attachment to the corolla wall is·close to the base in *Cobaea* and *Cantua*, low in the tube in *Bonplandia, Loeselia* and *Polemonium*, at a middle level in most of the genera, and in the sinuses of the lobes in some species of *Gilia, Ipomopsis, Eriastrum, Navarretia* and *Linanthus*. All the stamens of a set are inserted on the same level in most members of the family, but on three more or less widely spaced levels in the genus *Phlox* and in some species of *Collomia, Gilia* and *Ipomopsis*.

The filaments vary greatly in length. One extreme is represented by *Cobaèa penduliflora* with filaments 12 cm. long, the other by *Gymnosteris, Navarretia mellita*, and some species of *Phlox* with sessile or subsessile stamens. The filaments of a flower may be equal or somewhat unequal in length; both conditions are common.

The anthers are oval in most genera, linear in *Cobaea* and *Cantua*, and sagittate in *Eriastrum*. They are held in an erect position on the filaments in most members, but are versatile in attachment in *Cobaea* and *Eriastrum*. The anthers consist of two cells. Dehiscence is by longitudinal slits in each cell and is followed by shrinking of the anther wall tissue which has the effect of partly extruding the masses of pollen.

As regards the gynoecium the basic condition for the family is a superior three-celled ovary with a single style and three-lobed stigma. As already indicated, however, two-celled ovaries and two-lobed stigmas are found in certain species of *Navarretia* and *Leptodactylon*. The stigma is normally papillose. The nectary is a greenish disc composed of densely packed cells surrounding the base of the ovary. It is often conspicuously five-lobed with the lobes standing opposite the petals.

The vascular anatomy of the whole flower has been investigated in *Cantua buxifolia, Cobaea scandens, Phlox maculata, P. amoena, Linanthus breviculus, Navarretia breweri* and *N. pubescens* by Dawson (1936). The venation of the corolla was studied in *Navarretia* by Crampton (1954) and Alva Grant (unpubl.) and in *Gilia, Ipomopsis*, and *Eriastrum* by A. Grant. The results so far obtained indicate that a systematic survey of venation patterns within the family would be a rewarding task.

In *Cantua buxifolia* the primary stele of the pedicel is an unbroken cylinder of vascular tissue. At the base of the flower ten large bundles diverge from this cylinder. Five of these become the sepal midribs and the alternating five branch after a short distance. One branch becomes the lateral traces of the sepals so that the sepal whorl possesses fifteen bundles, a median and two lateral traces for each of the five sepals. The other branch forms the petal midribs. This whorl of bundles gives off two additional branches. Near the base five traces run a short distance into the five lobes of the nectary disc, while at a higher level the stamen traces diverge from the corolla. The five stumps of vascular tissue which supply the basal disc are interpreted by Dawson, on the basis of their position and vasculation, as vestigial stamen traces.

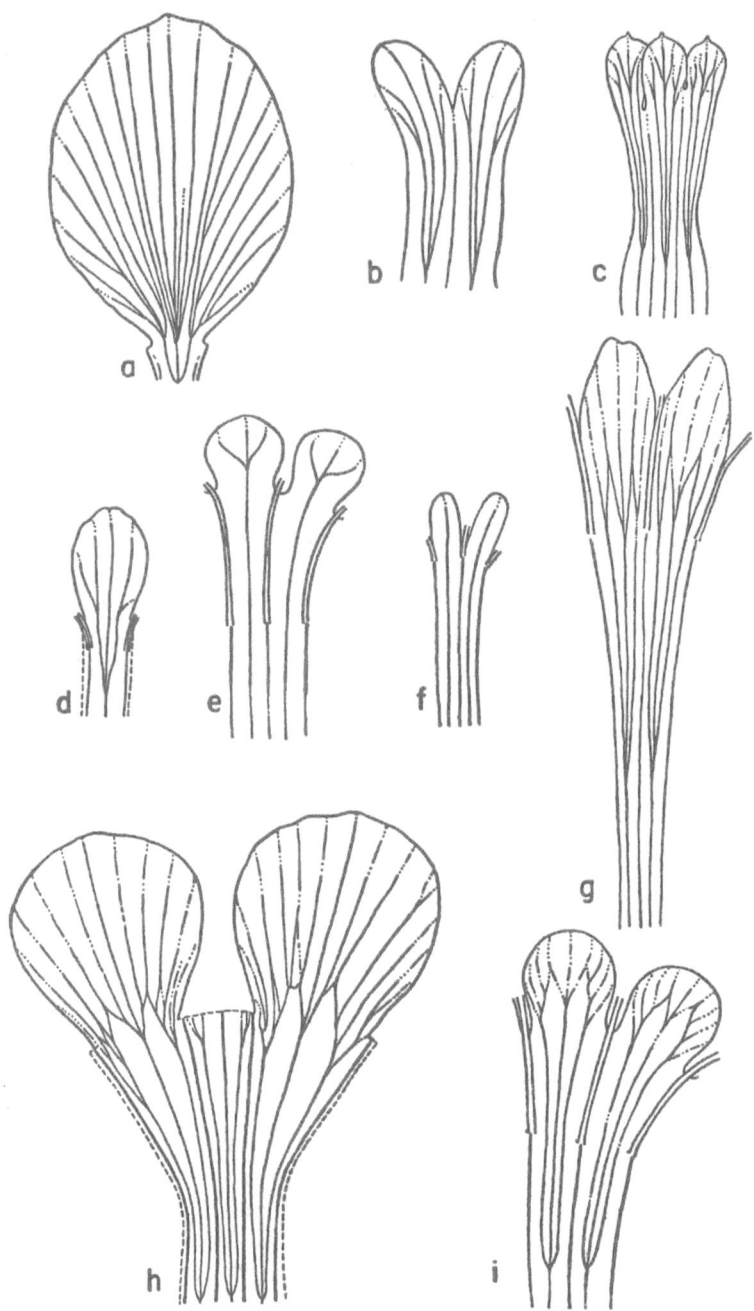

Fig. 14. Corolla venation.

a) Gilia rigidula. *b*) Ipomopsis congesta. *c*) Navarretia divaricata. *d*) Gilia tenerrima. *e*) Navarretia propinqua. *f*) Navarretia pleiantha. *g*) Gilia tricolor. *h*) Navarretia mitracarpa. *i*) Eriastrum
densifolium. (*e, f, h,* from Crampton; others original by A. Grant; magnified, not to same scale.)

Next in order of departure from the stele are the three midribs of the carpels, which promptly give off lateral dorsal branches. The remaining vascular tissue of the floral axis forms a compound bundle supplying the ventral walls of the three fused carpels. From this central bundle the traces to the ovules arise at a higher level. The dorsal, lateral and ventral bundles of the carpels continue through the style. All but the dorsal traces fade out in the upper part of the style and the dorsals alone enter the three stigma lobes.

Dawson found a high degree of uniformity in the vascular system of widely separated genera within the family. The vasculation of *Cobaea* and *Phlox* closely resembles that of *Cantua*, while *Linanthus* and *Navarretia* deviate due to a more extreme adnation and cohesion of parts. Thus in *Linanthus breviculus* there are ten and in *Navarretia breweri* five sepal traces instead of the fifteen found in *Cantua*. The petal and stamen traces in *Linanthus* depart from the stele at different levels rather than in regular whorls. Vestigial stamen bundles serving the basal disc are lacking in *Phlox*, *Linanthus* and *Navarretia*, with the exception of *Phlox amoena* in which they are very faintly developed, while in *Navarretia pubescens* not only the traces but also the disc itself is lacking. When compared with the prominent vascularized disc of *Cantua* and *Cobaea* these features indicate a more advanced state of reduction in the temperate herbaceous genera.

Several patterns of corolla venation are known in the temperate herbaceous genera. It is common to find three main veins running into each petal. These may diverge from a common bundle at the base of the petal (Fig. 14a–b) or at a higher level in the corolla (Fig. 14c–d). The primary branches may ramify further (Fig. 14a) or remain simple or nearly so (Fig. 14e–f). The primary branches may anastomose in the throat or lobes so as to enclose a pair of diamond-shaped areas (Fig. 14g). These areas may be broad (Fig. 14g–h) or so narrow as to be almost imperceptible (Fig. 14i), depending on the proportions of the corolla. One main vein to each petal alternating with a stamen bundle on the fused margins of adjacent petals represents still another condition (Fig. 14e). As stated earlier, the exploration of corolla venation in the Phlox family and the correlation of the venation patterns with taxonomic groups and phylogenetic trends is largely a task for the future.

THE GAMETOPHYTE

The ovules of *Polemonium caeruleum* are described by Sundar Rao (1940) as numerous, axile, anatropous, and possessed of a single massive integument with a somewhat curved micropyle. The nucellus is poorly developed, consisting of only a single layer of cells which in later stages become crushed so that the embryo-sac is directly surrounded by the integument. The inner epidermal cells of the integument become densely cytoplasmic and serve as a nutritive jacket for the embryo-sac. A similar condition was observed in *Collomia, Phlox, Gilia* and *Linanthus* by Billings (1901) and in *Cobaea scandens* by Dahlgren (1927: 400).

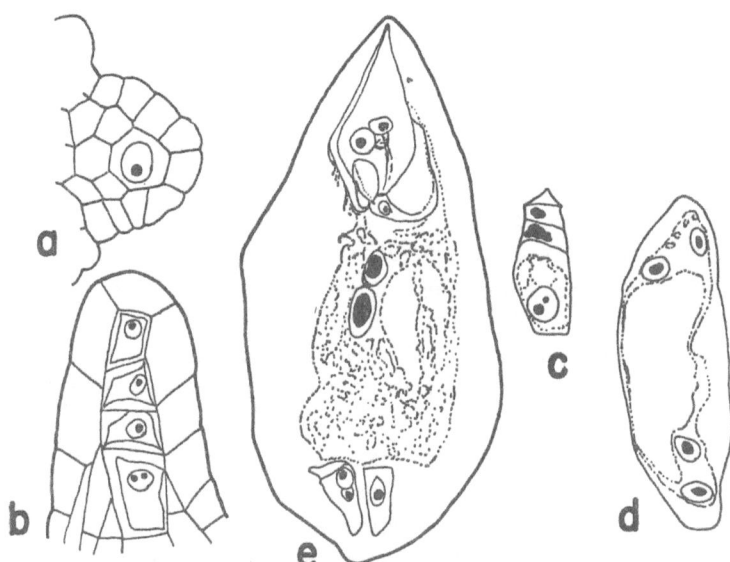

Fig. 15. Embryo-sac development in Polemonium caeruleum.
a) Primary archesporial cell in a young ovule. *b*) Tetrad of megaspores. *c*) Degeneration of the micropylar megaspores and growth of the chalazal one as a one-nucleate embryo-sac. *d*) Four-nucleate embryo-sac. *e*) Mature embryo-sac. (from Sundar Rao; × 533.)

The megaspore mother cell, derived from a specific hypodermal cell, gives rise to a linear tetrad of megaspores in both *Polemonium caeruleum* (Fig. 15a–b) and *Gilia millefoliata* (Sundar Rao 1940, Schnarf 1921). The three micropylar megaspores degenerate (Fig. 15c) and the chalazal one develops into an 8-nucleate embryo-sac as a result of three successive nuclear divisions (Fig. 15d–e). The embryo-sac is thus of the normal type in the investigated species of *Polemonium* and *Gilia*. The antipodals are small and degenerate at an early stage. At maturity the embryo-sac consists of a normal egg apparatus and two fused polar nuclei.

The following description of microsporogenesis in *Polemonium caeruleum* is taken from the account of Sundar Rao. Three or four rows of cells in each lobe of the young anther become differentiated as archesporial cells. Later each one of these cells cuts off a primary wall cell towards the outside. Divisions now proceed rapidly in both the wall cells and sporogenous cells while the anther attains its mature form. Successive periclinal divisions in the wall cells give rise to three layers, the inner one of which becomes the tapetum and the outer one a fibrous endothecium while the middle layer is crushed (Fig. 16a).

The first meiotic division of the pollen mother cells begins while the tapetal cells are in the resting condition. At about the stage of diplotene the tapetal nuclei divide without cell wall formation and the resulting binucleate cells persist until the end of meiosis. As the pollen mother cells reach the stage of cytokinesis the two nuclei in each tapetal cell fuse, restoring the uninucleate condition (Fig. 16b). The tapetum is now composed of tetraploid cells. The

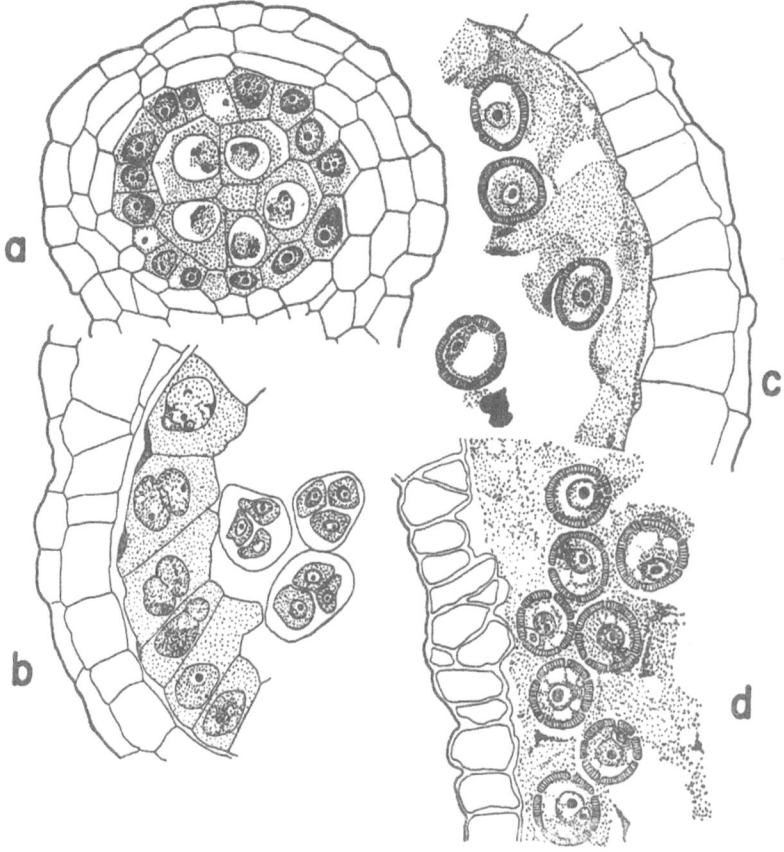

Fig. 16. Tapetum and pollen development in Polemonium caeruleum.
a) Young anther lobe showing primary archesporial cells in center surrounded by a tapetal layer (heavily stippled) and an anther wall composed of three layers of cells (unstippled). *b*) Part of an anther lobe showing much enlarged tapetal cells, fusion of the tapetal nuclei, and pollen tetrads. *c*) Uninucleate microspores and tapetum in an early stage of degeneration. *d*) Tapetum fully disintegrated and pollen grains binucleate. from Sundar Rao; *a, b,* × 466; *c. d,* × 266.

foregoing sequence appears to be true in *Gilia* as well as *Polemonium,* and in *Gilia* even octoploid tapetal cells have been seen (Fig. 17). Sundar Rao believes that Cooper's (1933) incomplete observations of tapetum development in *Phlox paniculata* may likewise conform to this pattern.

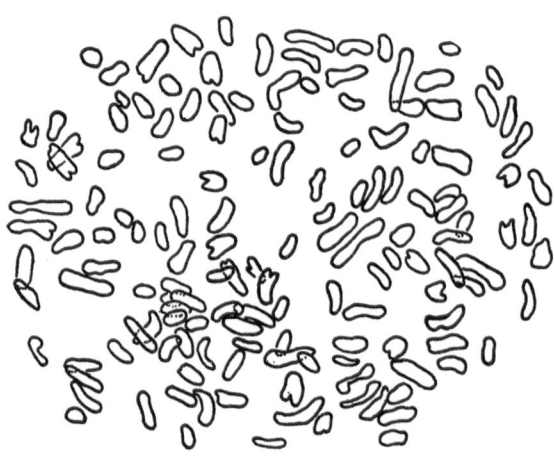

In *Polemonium caeruleum* and *Cobaea scandens* the tapetal cells enlarge and come to have thin walls when the microspores are in tetrads (Juel 1915, Sundar Rao 1940; see Fig. 16b). They lose their walls about the time that the microspores develop an exine and intine (Fig. 16c). The nuclei of the tapetum degenerate and the cytoplasm flows between the pollen grains when the pollen becomes binucleate (Fig. 16d). Ultimately this soupy mass of tapetal material, described as a periplasmodium in *Cobaea* and a false periplasmodium in *Polemonium,* is absorbed by the developing pollen grains.

Fig. 17. Division of a tapetal nucleus in Gilia transmontana. The somatic number for this species is 2n = 36. Approximately 148 chromosomes are visible in the camera lucida drawing of an early anaphase. This corresponds to about 74 chromosomes for each daughter nucleus as well as for the original mother nucleus. This tapetal cell was thus tetraploid through fusion of the nuclei in its first binucleate condition, and in its second binucleate condition would be octoploid. (original; × 1400.)

Cytokinesis in the tetrad of microspores takes place by simultaneous furrowing in *Polemonium caeruleum* and *Cobaea scandens* (Sundar Rao 1940, Lawson 1898, Farr 1920). Growth of the microspores after liberation from the tetrads involves a marked increase in size, development of an exine and intine, and formation of a large vacuole which pushes the nucleus to one side. Here the nucleus divides to form a large tube nucleus and a small generative nucleus. The generative nucleus becomes walled off by a curved wall so as to form a lenticular generative cell. The pollen grains are shed in the binucleate condition in *Polemonium, Gilia* and *Phlox* (Sundar Rao 1940, Schnarf 1937).

The external morphology of the pollen grains has been studied by Erdtman (1952: 328–331) and Alva Grant (unpubl.). The following resumé is based on manuscript notes of the latter student. The pollen grains of the Polemoniaceae are slightly flattened spheres with pores in the exine. Among the important differences between the genera and species are size of the grains, number and distribution of the pores, structure of the pores, and thickness and structure of the exine.

The size of the pollen grains varies widely within the family. In general the species with large plant bodies and large flowers have larger grains than those with small plant bodies and small flowers. Thus *Cobaea scandens*, a large woody vine with big flowers, has pollen about 150 microns in diameter, whereas *Gilia campanulata*, a diminutive annual with small flowers, has pollen 25–34 microns in diameter. Within the section *Arachnion* of *Gilia*, considered as a whole, the pollen diameters vary from 40 to 75 microns. Pollen grains less than 52 microns in diameter are found only among the species with the smallest bodies and smallest flowers; pollen greater than 61 microns in diameter is (with the exception of a few small-bodied tetraploid species) found only among the species with the largest bodies and largest flowers.

On the other hand, three species of *Gilia* sect. *Giliastrum* have evolved minute pollen regardless of plant size. Thus *Gilia latifolia*, a relatively large annual, produces pollen 30 to 40 microns in diameter. The genus *Linanthus* has also developed fairly large-bodied species with small pollen grains.

The number of pores on a pollen grain ranges from numerous to few. The minimal numbers of pores so far found are five and four. These pores may be evenly spaced over the entire surface of the grain or arranged around the equator. The equatorial arrangement is found only in species with few (4 to 12) pores. Several species exhibit a transitional condition between a uniform and an equatorial distribution of the pores. *Gilia capitata* and *G. tricolor*, for example, have an imperfect, nearly equatorial pore arrangement.

Two layers may be recognized in the exine, an outer layer which is usually thicker, and a thin inner exine. A pore is a round or oval aperture in the inner exine surmounted by an opening of a similar or different shape in the outer exine. The gap in the outer exine may correspond in size to the opening in the inner exine, or it may be smaller and thus present an overhanging edge. A common type of pore found where the pores are equatorially arranged is the combination of a circular opening in the inner exine and an elliptical opening in the outer exine.

The outer exine consists of protuberances in the forms of rods, ridges, or combinations of these elements which in surface view appear as dotted, striated or reticulate patterns, or it consists of scattered knobs or warts. These structural patterns are well developed in most species of Polemoniaceae, but may be weakly developed in other cases (viz., *Gilia rigidula*) where the exine is unusually thin. Some species and perhaps whole genera are characterized by a heavy exine.

A simple pattern is the uniform distribution of small dots formed by short projecting rods over the outer exine. The beginning of a more complex exine pattern is found where the dots are arranged in short dotted lines. These lines may then be oriented with respect to one another in either a series of striations or a reticulation. Well developed exine patterns may consist of dotted lines, solid lines, or a combination of both. The increasing complexity of the exine pattern follows several different trends within the family.

The nature of the exine pattern is closely related to the number and arrangement of the pores. In species with many closely spaced pores a reticulate arrangement of the ridges of the outer exine provides in a neat way the outer aperture for each pore. Thus in *Phlox* and *Cobaea* each pore is surrounded by a polygonal "fence" constructed in this manner. Erdtman (1952) reports on certain *Phlox* species in which some of the meshes of the reticulum have apertures and hence open pores and some do not. The exine pattern in *Microsteris gracilis* he describes as being made up of rhombus-shaped units containing nine meshes each, of which four surround open pores and five are not associated with pores.

Striate patterns are often found in species having few, well-spaced pores. The lines have a flowing or swirling arrangement in *Eriastrum diffusum*. Where the pores are equatorial in position the lines may be oriented in the manner of a magnetic field (*Langloisia punctata, Gilia latiflora*). The ridges of the outer exine often separate or fork above the pore opening in the inner exine so as to partly cover over the pore on two edges and give it the elliptical surface configuration described above. In *Collomia linearis*, by contrast, the ridges break off at each pore, permitting its full exposure.

Resembling somewhat the simple dotted surface first described, but specialized in a different way from the striate or reticulate patterns, is a rugose type of pollen grain in which the protuberances have become large and knobby or warty. In *Cantua candelilla* the surface is uniformly covered with roundish angular processes except for scattered clear areas in which the pores occur. A similar type of outer exine is found in *Huthia*.

Germination of the pollen grain, growth of the pollen tube and fertilization take place in 12 to 36 hours in *Phlox drummondii* (Miller and Wetmore 1945a).

FRUIT AND SEED

The enlargement of the fruit in *Cantua, Huthia, Bonplandia* and *Polemonium* is accompanied by enlargement of the tubular but accrescent calyx. The herbaceous calyx of *Collomia* with its plaits in the sinuses partly grows and partly unfolds with the expansion of the fruit. The large fruit of *Cobaea* simply pushes apart the free sepals. In the genera with membranous calyx sinuses (*Allophyllum, Phlox, Gilia, Navarretia, Linanthus*, etc.) the calyx becomes split along these weak points by the growth of the fruit. In *Gilia* the calyx enlarges in fruit before it eventually ruptures. In the Phloxes with carinate sinuses the moment of splitting is postponed by the unfolding of the sinus. The calyx of *Phlox drummondii* is resistant to the pressure of the growing fruit, causing the latter to dehisce and forcibly eject the seeds.

The fruit is a capsule throughout the family. It is normally three-celled but in some species of *Navarretia* is two- or one-celled. It dehisces at maturity along the back of the carpels in all genera except *Cobaea*, where dehiscence occurs on the common septa of adjacent carpels. The valves normally open from the top downwards (Fig. 18a), but in some species of *Navarretia* open from the base upwards. Circumscissile dehiscence is present in the section *Mitracarpium* of *Navarretia*. Indehiscent capsules are found in *Navarretia* sect. *Navarretia* and *Gilia* (Fig. 18b).

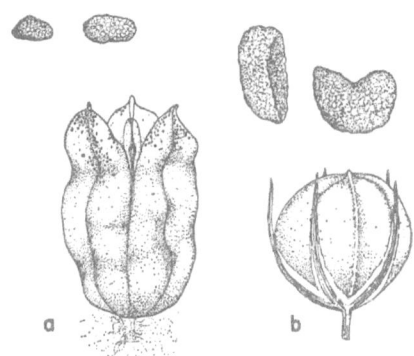

Fig. 18. Capsule and seeds in Gilia capitata. a) Dehiscent capsule and fine seeds of Gilia capitata chamissonis. b) Indehiscent capsule and large seeds of Gilia capitata capitata. (from Grant; × 6.)

The capsule wall is normally of a hard or leathery texture but is thin and membranous in certain members of *Navarretia*. The indehiscent, single-celled, membranous fruits of the vernal-pool inhabiting Navarretias are, as Brand pointed out, more nearly berries than capsules. The wide range of fruit types found within the genus *Navarretia*, which was brought to the attention of botanists by Greene, has been reviewed recently by Crampton (1954).

The mature fruits are in most cases many-seeded. The number of seeds is reduced to one or two per cell in *Collomia*, *Allophyllum*, *Phlox*, *Microsteris*, and species of *Gilia*, *Ipomopsis*, *Eriastrum*, *Navarretia* and *Linanthus*. Those plants which have indehiscent capsules also have few seeds (Fig. 18). Single-seeded dehiscent capsules are characteristic of the cespitose Phloxes and two- or three-seeded indehiscent capsules are found in *Navarretia*. The ultimate reduction to single-seeded indehiscent capsules has been attained by *Navarretia pauciflora* and *Gilia capitata*. The seeds formed in many-seeded capsules are usually small in size and angular and irregular in shape, whereas the solitary seeds are larger and more rounded. The single-seeded fruits may develop from ovaries in which numerous ovules are present in each cell (*Gilia capitata*) or in which the number of ovules is also reduced to one per cell (the cespitose Phloxes).

The seeds of the Phlox family range in size from objects nearly 2 cm. long in *Cobaea* to fine dust-like particles in many desert species of *Gilia*, *Ipomopsis*,

Linanthus and *Gymnosteris*. In general, the tropical woody Polemoniaceae have large to medium-sized seeds, while medium-sized to minute seeds are found in the temperate herbaceous genera. The larger size of the tropical seeds is due to a flat bordering wing, which is broad in *Cobaea*, *Cantua* and *Huthia*, narrower in *Loeselia*, and very narrow in *Bonplandia*. Vestiges of this wing can be seen in the seeds of a few species of *Polemonium*. Otherwise all the temperate Polemoniaceae have plump unwinged seeds.

The seed-coat presents some interesting features. According to Brand (1907: 12) four conditions are found: (i) an outer layer of the coat is separable from an inner layer, and the outer layer remains unchanged when wet; (ii) two separable layers are present as before, but the outer layer becomes mucilaginous when wet; (iii) the outer and inner layers of the coat are tightly fused, and the outer layer is unchanged by water; (iv) the two layers are fused as before, but the outer layer becomes mucilaginous when wet. Condition (iii) is found in *Phlox* and most species of *Polemonium*, and condition (iv) in *Collomia*, *Microsteris* and many species of *Gilia*. The first two conditions are more uncommon, (i) being found in *Linanthus parryae*, *L. demissus*, *L. dichotomus*, and *Ipomopsis rubra*, and (ii) in *Allophyllum* and *Gilia patagonica*.

The mucilage of the seed coat may arise from coiled processes or spiracles attached to the outer surface or from a combination of spiracles and slime cells. The outer coat when wet is swollen to almost double its normal size in *Collomia*, is moderately swollen in *Gilia*, and is scarcely changed in size in *Gymnosteris*. The histological nature of the seed coat deserves a thorough comparative study.

The seeds in most genera are well supplied with nuclear endosperm (Schnarf 1929: 378). According to Brand (1907: 13) the endosperm is scanty in *Cobaea* and *Cantua*, and almost completely wanting in *Gilia subnuda*, *Linanthus breviculus*, and *Navarretia atractyloides*.

CHAPTER III

PROBLEMS OF CLASSIFICATION

The Polemoniaceae is a taxonomically complex family in which generic and specific divisions have frequently been the subject of confusion and dispute. The situation was well stated by Dawson in 1936. "The Polemoniaceae, comprising [eighteen] genera and about three hundred species, while clearly defined as a natural order of plants, show unusual difficulties in the determination of generic relationships. The entire group is so closely interrelated that absolute characters, which could serve as a basis to separate genera and species taxonomically, are lacking. Since the time of Jussieu (1789) this family has presented an interesting but baffling problem for systematists. Authors have differed widely in their definition of genera. This uncertainty of status is reflected in the confusion of nomenclature existing today, as well as in the wide variations of opinions regarding the merging or segregation of genera and species."

Professor Wherry has noted the appropriateness of naming the family after an ancient philosopher, Polemon, whose name means The Fighter. Fortunately, polemoniaceous taxonomists during the past generation have devoted themselves to some sound research as well as to polemics, and as a consequence the latest brief generalized classification of the family by Peter (1891) and the latest monographic treatment by Brand (1907) are now very much out of date. The traditional generic alignments correspond only moderately well with the main phylogenetic trends as indicated by much recently obtained morphological and distributional evidence.

The authors of the successive general treatments of the Phlox family, Bentham, Gray, Peter, Brand, lived and worked in Europe or eastern North America remote from the center of distribution of the family. Their knowledge of the plants was of necessity based mainly or even entirely on herbarium specimens and garden-grown materials. Most of these authors never saw the plants in their native habitats and none of them had the benefit of a wide field experience with the family. It is not surprising that a half century after the latest of the general treatments it should be necessary to go over the system of classification again and introduce a number of modifications.

One of the main purposes of the present study is to redefine and circumscribe the tribes, genera and sections of the Phlox family in the light of present knowledge. The revised system of classification is discussed in this chapter and presented formally in Chapter 4, while the nomenclatural details are attended to in Chapter 5.

PRIMARY CLASSIFICATION

The tropical genera of the Phlox family, *Cobaea, Cantua, Huthia, Bonplandia* and *Loeselia,* possess a number of common characteristics. They are woody and leafy plants with large corollas possessing a supernumerary whorl of stamen traces, flat winged seeds, colorless embryos, and minute chromosomes. These genera also differ among themselves in important respects and are separated in this and other treatments into three tribes. The distinguishing characteristics of the genera are summarized in Table 2 and those of the tribes in Chapter 4.[1]

The first of the tropical tribes is the Cobaeeae. Opinions concerning the systematic position of the genus *Cobaea* have varied between wide extremes, as exemplified on the one hand by Don (1824) who isolated it in a family by itself and on the other hand by Reichenbach (1837) and Wherry (1940) who joined it with *Polemonium* in a special tribe. Most authors (Meisner 1839, Baillon 1890, Peter 1891, Brand 1907) have regarded this genus as belonging to the Polemoniaceae but sufficiently different from the other members to warrant separation in a tribe of its own. My own estimate of the situation based on the summary of morphological characters given in the next chapter is in agreement with the classical disposition. Nevertheless there may be a strong element of truth in Wherry's and Reichenbach's postulation of a special relationship between *Cobaea* and *Polemonium.* We will return to the latter point in a later discussion.

The woody Andean genus *Cantua* was also placed by Peter and Brand in a separate tribe (cf. Table 1). Wherry on the other hand suggested that its characters "may be the result of evolutionary advance from *Collomia,*" and on the basis mainly of leaf and calyx features grouped it with *Collomia, Bonplandia, Gymnosteris* and *Navarretia* in a "Collomia tribe". It seems to me that this latter arrangement can only be maintained by ignoring other important characters which differentiate these genera and point to relationships in other directions. Here again the morphological evidence summarized below seems to me to favor the traditional treatment involving a separate Cantua

[1] For a complete listing of the characters throughout this discussion refer to Chapter 4.

tribe. *Huthia* is related to *Cantua* by its calyx, corolla, fruit and seed characters and its geographical distribution and falls naturally into the Cantua tribe. As indicated in the introductory statement there are significant resemblances between the Cantueae and *Cobaea*.

With respect to the genera *Loeselia* and *Bonplandia*, Wherry's suggestion (op. cit.) that *Loeselia* is allied to *Phlox* and *Bonplandia* to *Collomia* and hence belong in separate tribes is not the only or even the most likely conclusion that can be drawn from the morphological evidence. The general habit, axillary disposition of the flowers, slightly zygomorphic calyx and corolla, and presence of wings on the seeds are much alike in *Bonplandia* and *Loeselia*. These morphological likenesses, as well as numerous other characters differentiating them collectively from the other Polemoniaceae, support Brand's assignment of the two genera to a separate tribe Bonplandieae.

The vast majority of the Polemoniaceae are plants of the temperate zone. In contrast with the tropical genera they are usually herbaceous, though sometimes woody, plants; some or all of the leaves are frequently reduced in size; the corolla is generally smaller with only one whorl of stamen traces; the seeds are plump and not winged; the embryo is usually green; and the chromosomes are larger in size.

Most systems of classification (Baillon 1890, Peter 1891, Brand 1907, but not Wherry 1940, cf. Table 1) have grouped the temperate and boreal genera into one large tribe, the Polemonieae. It seems to me that this grouping is both cumbersome and unnatural. Two main lines of evolution can be discerned within the temperate herbaceous Polemoniaceae. On the one hand the genera *Polemonium*, *Allophyllum*, *Collomia*, *Gymnosteris*, *Phlox* and *Microsteris* form a natural assemblage, which Grant and Grant (1955) have called "the Collomia phylad" and which we can now treat formally as the Polemonieae sensu stricto. The remaining North American genera, *Gilia*, *Ipomopsis*, *Eriastrum*, *Langloisia*, *Navarretia*, *Leptodactylon* and *Linanthus*, form another natural assemblage, which may be designated by the old name Gilieae of Reichenbach.

The Polemonieae and the Gilieae differ more in general tendencies than in sharp distinctions, which is why they have not been set apart as tribes in the past. The Polemonieae are predominantly perennial herbs with leaves of simple or no dissection (Figs. 10, 11a–c); the calyx is usually herbaceous; the corolla regular; and the point of insertion of the stamens, if uneven at all, is usually very strongly so. By contrast the Gilieae are predominantly annuals though including many perennial types; the leaf form may be simple but if dissected is apt to be cut in a complex fashion (Figs. 11d–g, 12); the calyx is differentiated into herbaceous lobes joined by membranous tissue in the sinuses; the corolla is often zygomorphic; and the stamen insertion is usually even or at least not strongly uneven. The Polemonium tribe has its center of distribution in cold temperate regions in North America and is the only tribe represented in the arctic and Eurasia, whereas the Gilia tribe centers in the arid southwestern region of North America and is the only tribe represented in the hot deserts.

The separation of the Polemonieae and Gilieae as distinct tribes cannot be defended on the basis of convenience in classification. No single key character is known which will separate the two series in every particular instance. The main argument in favor the segregation of two tribes in the herbaceous temperate Polemoniaceae rests rather on the clear evidence of special relationships among the genera composing each tribe.

Within the Polemonium tribe, *Allophyllum* is close to *Polemonium* and *Collomia*; *Gymnosteris* is close to *Collomia*; and *Microsteris* to *Phlox*. *Collomia* is more advanced in nearly every character than *Polemonium*, but the derivation of both lines from some common ancestor presents no serious difficulties. *Phlox* is more advanced than *Collomia* but the gap between the two genera is not especially large and it is probable that they have diverged from a common line. In one way or another, *Polemonium*, *Allophyllum*, *Collomia*, *Gymnosteris*, *Phlox* and *Microsteris* are interrelated and should, consequently, be grouped in a common taxonomic category.

In the tribe Gilieae, all the genera are closely enough interrelated to have been placed in a single genus at one time or another. As in the Polemonieae, certain subtribal groupings are evident. The genus *Gilia* as construed in this work is well differentiated from the rest of the tribe by its scapose or semiscapose habit, spring season of flowering, regular funnelform corolla, and 9 pairs of large chromosomes. There are exceptions to each one of these distinguishing characters. *Ipomopsis*, *Eriastrum*, *Langloisia* and *Navarretia* have enough common features to be regarded as belonging to a separate circle of affinity. These common features include a summer blooming habit, strongly developed upper leaves and frequently leafless lower stems, a tendency to have bristle-tipped or prickly leaf segments, bracts and calyx lobes, the frequent occurrence of bilabiate corollas, and as a general rule 7 pairs of chromosomes. *Leptodactylon* and *Linanthus*, finally, share in the possession of palmately divided leaves, spring blooming season, regular and often salverform corollas, and 9 pairs of small chromosomes. These two genera evidently form a third circle of affinity. The degree of distinctness of the three subtribal alliances in the Gilieae is on a par with that of the *Polemonium-Allophyllum*, the *Collomia-Gymnosteris*, and the *Phlox-Microsteris* groupings in the Polemonieae.

The Phlox family can be subdivided on the basis of a combination of characters into two main groups, the tropical genera and the temperate genera, as shown above and also in Chapter 4. Brand (1907) treated the two large divisions of the family as subfamilies (cf. Table 1), and it might seem logical to follow this arrangement. In my opinion, however, the Cobaeoideae and Polemonioideae of Brand, as well as the differently composed Tropical and Temperate divisions outlined in this work, represent artificial rather than natural groups. The latter grouping is retained here solely for purposes of convenience.

The indication of similarities and relationships between the tropical genera is not the whole story. There are also wide differences between these genera and, on the other hand, certain indications of relationship between particular tropical and temperate genera, which tend to be glossed over in attempts to maintain a formal division into two subfamilies.

Similarly the temperate genera fall into two rather remotely related series which seem to be connected independently of one another with different tropical members. The large gaps surrounding the Polemonieae and Gilieae narrow down in different directions, suggesting a remote connection of the Polemonieae with *Cobaea* and of the Gilieae with *Loeselia*. There is no convincing evidence, at any rate, of a close connection between the Polemonieae and Gilieae. For this reason any collective tribe or subfamily defined so as to include all the temperate genera is an artificial taxon from the standpoint of phylogenetic taxonomy.

If we consider the differences as well as the similarities within the family we are bound to recognize the existence of various highly isolated phylads. The interrelationships between these phylads, as suggested by the morpholoical characters and geographical distribution, are loose and more or less reticulate, rather than simple and dichotomous. The Cobaeeae are apparently related in some respects with the Polemonieae, in other respects with the Bonplandieae, and in still other respects with the Cantueae. The Bonplandieae are believed to be related with the Gilieae as well as with the Cobaeeae. This reticulate pattern of relationships between strongly isolated groups is expressed more adequately by a division into five than into two major categories.

THE GENUS PROBLEM

Genera have been circumscribed in this work in accordance with the following criteria. It is assumed, in the first place, that the genus as a natural unit must correspond to a series of populations related by descent from a common ancestor. Its scope is prescribed by its position in the hierarchy of taxonomic categories between section and tribe. Genera, considered as main phyletic branches, should not normally be capable of crossing with one another, and this qualification sets a lower limit on the scope of this entity. Its upper limit within the tribe is determined by manageability and convenience in classification. I personally prefer to define the genus as the most comprehensive natural infra-tribal category which can be handled practicably in the field or herbarium, rather than as the minimum distinguishable supra-specific category, because it is to the advantage of everyone who uses generic names to have as few of them to remember as possible.

Naturalness, phylogenetic relationship, and complete genetic isolation are objective phenomena. Within the broadest limits the genus can be described in objective terms. "Manageability", "practicability", and "convenience," however, are relative terms, so that in a precise sense the genus cannot be defined objectively. In the final analysis, to paraphrase a famous dictum of Darwin, "the opinion of naturalists having sound judgment and wide experience seems the only guide to follow."

The historical experience of students of the Polemoniaceae has provided a partial solution to the question of what constitutes a practicable genus definition in this family. The attempt of Gray to reduce all of the temperate genera to *Phlox*, *Polemonium* and *Gilia*, and of Kuntze to reduce these three still further to *Polemonium*, has been rejected unanimously by subsequent students. It was rejected because the resulting *"Gilia"* or *"Polemonium"* came to include too many diverse groups to be meaningful as generic names. There is no advantage in having the category of genus at all in a diversified family if the genus is virtually synonymous with the family.

Gray was led to the wide-scale lumping of the temperate Polemoniaceae by the apparent failure of key characters in exceptional instances. The difficulty, however, lay not in the genera themselves, but in Gray's assumption that they could be satisfactorily characterized by single features rather than by combinations of characters. This point has been discussed by Mason.

"The predominantly annual species of Polemoniaceae can be divided into natural genera and *Eriastrum* is one of them. In the past, great weight has been placed upon certain key characters in the differentiation of the genera of this group of plants. Use of a particular character has often been inherited from the keys of our predecessors and may date back to early beginnings when only a few species were known or in some cases even from times when the subgenera and genera were very unnatural. Such key characters are often erroneous, as is the stamen character most frequently used to separate *Eriastrum* from *Navarretia* ... Despite the invalidity of this 'key' character, these two genera are none the less distinct from one another. Genera do not stand or fall solely on good or bad key characters ... We are indeed fortunate when differences can be stated in precise terms of single characters but differences are none the less important when they must be grouped to give character to the whole" (Mason 1945).

What is to prevent us from going to the opposite extreme of promoting each section to the status of a genus? This has also been tried and found unsatisfactory. Most of the present sections of *Linanthus* were treated as genera in the time of Bentham and Nuttall. Rydberg treated several of the present sections of *Gilia* as genera. Neither author is followed today. Experience has shown that the main phyletic branches within a tribe can be hidden from view as successfully by assigning too many generic names to them as by assigning too few.

The application of a reasonable and scientifically useful genus concept in the Polemoniaceae has met with difficulty in only a few sectors of the family. There has never been any serious disagreement over the limits of the tropical genera, *Cobaea*, *Cantua*, *Huthia* and *Bonplandia*. *Polemonium* and *Phlox* were recognized as distinctive groups in pre-Linnean times and, with the exception of one ambitious but abortive plan to lump them, have enjoyed an unbroken acceptance as genera ever since.

The greatest difficulty of generic delimitation in the family has centered on *Gilia*. As handed down by Gray this genus was a hodgepodge of dissimilar and unrelated elements. Wherry (1940) has observed that, as the traditional catch-all for the family, any new species of uncertain affinities was automatically described as a *Gilia*. All but 19 of the temperate species of Polemoniaceae outside of *Polemonium* and *Phlox* have been in *Gilia* at one time in their taxonomic history.

The exclusion by Greene of *Collomia*, *Gymnosteris*, *Microsteris*, *Langloisia*, *Navarretia*, and *Linanthus* (Greene 1887, 1892, 1896, 1898); by Jepson of *Eriastrum* and *Leptodactylon* (Jepson 1925); and by the Grants of *Allophyllum* and *Ipomopsis* (Grant and Grant 1955, Grant 1956), leaves the genus a somewhat more manageable and comprehensible unit. It remains, nevertheless, after all these segregations the second largest genus in the family, and is still the most heterogeneous.

Eriastrum, which was formerly treated as a subdivision of *Gilia*, then of *Navarretia*, was raised to generic status by Jepson (1925) under the name *Hugelia*. Wheeler (1942) and Mason (1945) independently called attention to the fact that the oldest available name in the rank of genus was *Eriastrum*. The necessary transfers to the valid generic name were made by Mason in 1945.

The distinctiveness of the *Ipomopsis rubra-aggregata* group from *Gilia* proper has been noted by numerous botanists, more specifically by Pursh, Nuttall, Michaux, Lindley, Rafinesque, Greene, Flory, Wherry, and Grant. A nomenclatural change which effected the segregation of the eastern American species, *Ipomopsis rubra*, from *Gilia* was made by Wherry in 1936, but this assignment was not followed by Fernald or Gleason in the 1950 and 1952 editions of their Floras.

The genus *Ipomopsis* was subsequently construed by Grant (1956) in a broader sense than by previous authors, through inclusion of the old *Gilia congesta* and *Gilia multiflora* groups. The morphological and ecological resemblances between the latter groups and the *Ipomopsis aggregata* group suggested that the line of separation between *Gilia* and *Ipomopsis* should be drawn so as to include these three groups in one genus. The differences between them were regarded as being worthy of sectional status. *Ipomopsis* has more characteristics in common with *Eriastrum* than with *Gilia*.

Brand (1907) divided the genus *Loeselia* into two sections, *Giliopsis* and *Euloeselia*, the former containing four species, *L. tenuifolia, guttata, havardi,* and *effusa*. Grant (1956) regarded *Loeselia tenuifolia* as a member of *Ipomopsis* sect. *Phloganthea,* Gray noted in the original description of *L. guttata* that it is closely related to *L. tenuifolia;* and Kearney and Peebles (1943) placed *L. havardi* in the *Gilia multiflora* group which in Grant's classification is also *Ipomopsis* sect. *Phloganthea*. The fourth species of *Loeselia* sect. *Giliopsis*, *L. effusa*, belongs in *Gilia* sect. *Giliastrum*. It thus appears that only the members of the section *Euloeselia* of Brand can be maintained in the genus *Loeselia*.

The taxonomic history of *Linanthus* differs in a curious way from that of *Gilia*. In the latter the demands of naturalness have been met by splitting off various unrelated or distantly related series. In *Linanthus*, by contrast, phylogenetic unity has been attained by bringing together several originally separate genera, such as *Linanthus* sensu stricto, *Leptosiphon, Fenzlia,* and *Dactylophyllum*. The group most recently brought into the fold is *Linanthastrum*, which seems fairly close to *Linanthus grandiflorus*. The features which bring the various groups together in *Linanthus* and set them apart from the rest of the family are numerous, but of primary importance are the herbaceous habit and opposite palmate leaves.

When the family Polemoniaceae is considered as a whole many instances are found of minor genera attached to major genera. *Huthia* is thus related to *Cantua, Gymnosteris* to *Collomia, Allophyllum* to *Polemonium* and/or *Collomia, Langloisia* to *Eriastrum,* and *Leptodactylon* to *Linanthus*. No recent student has proposed to reduce any of these minor genera and indeed nothing would be gained by such a reduction. The minor genera are amply distinct from their major counterparts. Under these circumstances the working taxonomist must formulate a pair of group concepts in each case anyway, and he can easily associate the two concepts with two generic names.

Some authors have gone further in the segregation of minor genera which are not so clearly set apart from their main body of relatives. *Polemoniella* has been separated from *Polemonium* (Heller 1904, Wherry 1940, 1944a); *Linanthastrum* from *Linanthus* (Heller 1912, Ewan 1942, Jepson 1943, Wherry 1945b); and *Microsteris* from *Phlox* (Greene 1898, Wherry 1940, 1943, 1955).

These segregations have been opposed by other equally competent students on the grounds that they obscure rather than illuminate the relationships (Jepson 1943 and Davidson 1947 re *Polemoniella;* Mason 1951 re *Linanthastrum;* Mason 1941, 1950 re *Microsteris*).

The issue is unlikely to be settled in any definite way since its resolution falls wholly within the realm of the taxonomic judgement of different authors. In the present work I have preferred to keep *Polemoniella* in *Polemonium* and *Linanthastrum* in *Linanthus* because of the closeness of the relationship and the weakness, as I see it, of the case for splitting. After some debate I have also decided in favor of the segregation of *Microsteris*.

This latter assemblage has been placed in synonymy with nearly every large temperate genus except *Linanthus*. It is now generally agreed to be allied as a reduced derivative to *Phlox*. The differences between *Phlox* and *Microsteris* are noteworthy and correspond well with the pattern of major genus and satellite found throughout the family. In the case of *Microsteris* the balancing of pros and cons thus leans toward generic segregation.

INFRAGENERIC CLASSIFICATION

The only infrageneric category employed in the present treatment is the section. Certainly some sections of a given large genus are more closely related than others and could be grouped into a subgenus. Particular species within a section likewise frequently form a natural group of their own, which could be treated as a subsection or series. It is never possible, however, to express all the fine differences in degree of relationship by means of fixed categories in the taxonomic structure. The introduction of additional categories eventually reaches a point of diminishing returns beyond which the gain achieved by 'a finer expression of the interrelationships is ·outweighed by the loss due to purely structural complications in the system of classification. The taxonomist must beware of ensnaring himself in a taxonomic structure of his own creation.

The three sections of *Cobaea* recognized in this work are those of Brand (1907). The three subdivisions of *Polemonium* are essentially as drawn up by Peter (1891) and reinterpreted by Brand (1907) and Wherry (1942). The three sectional groupings in *Collomia* and *Phlox* are due to Wherry (1944b, 1955). The infrageneric classification of *Phlox* proposed by Wherry in 1955 is very different from earlier arrangements adopted by Wherry himself and by Brand. I have taken up the earliest sectional names for Wherry's sections in accordance with the rule of priority. *Ipomopsis* was subdivided into three sections by Grant (1956) and this classification is followed here.

With regard to *Loeselia*, I am setting the very distinctive *L. grandiflora* apart in a section separate from the other species of the genus. When *Loeselia* in general and *L. grandiflora* in particular are known better it may prove desirable to remove this latter interesting and distinctive plant to a genus of its own.

There have been three attempts prior to the present one during this century to divide the genus *Gilia* into subgenera and sections (Milliken 1904, Brand 1907, Mason and A. Grant 1948). None of the previous classifications has been based on as much information as is now available. Milliken proposed eight subgenera which are roughly equivalent to the present genera *Allophyllum*, *Microsteris*, *Gilia*, *Ipomopsis*, *Eriastrum*, *Leptodactylon*, and *Langloisia*. In Brand's system *Gilia* was divided into three subgenera, *Benthamiophila*, *Grayophila* and *Greeneophila*, each of which was further subdivided into sections. The "Bentham-loving Gilias" comprised four sections of which one corresponded roughly to the *Gilia* of the present treatment, two corresponded to the present genus *Ipomopsis*, and the fourth contained elements now placed in *Microsteris*, *Allophyllum*, *Ipomopsis* and *Gilia*. Brand's "Gray-loving Gilias" are *Linanthus* and his subgenus *Greeneophila* remains in *Gilia*.

Against this background Mason and A. Grant's system must be regarded as a notable advance in *Gilia* taxonomy. Of the extraneous elements only *Allophyllum* and *Ipomopsis* were retained in *Gilia* and the distinctiveness of these was recognized. The treatment of the true Gilias, however, suffered from a preoccupation with the Pacific coast forms so that the classification falls down in the Rocky Mountain region and the Southwest. Also the relationships of the species now composing the section *Saltugilia* were not yet understood in 1948. The section now bearing the name *Arachnion* was construed too broadly in the earlier classification and the section *Gilmania* too narrowly. The classification adopted here, which is a product of the work of Grant and Grant, represents an attempt to block out more natural sections in *Gilia*. Five sections are recognized.

The genus *Eriastrum* is fairly uniform, and while species groups can be discerned, such as the *Eriastrum densifolium-pluriflorum-eremicum* group, these do not seem to be of sectional importance.

A comparative morphological survey of corolla venation, stamen insertion, capsule dehiscence, vessel anatomy and other features in the genus *Navarretia* was made by Crampton (1954). On the basis of these studies Crampton was able to show the derived nature of the species which occupy specialized vernal pool habitats. On morphological and ecological characteristics he set apart the most reduced species as a special section, which except for the exclusion of two species, is accepted here. Unfortunately nomenclatural technicalities stand in the way of adopting Crampton's name for this section. The remaining species, which were placed by Crampton in one other section, form a heterogeneous assemblage which is here treated as three sections, making a total of four sections for the genus as a whole.

The latest author to deal with infrageneric divisions in *Linanthus* was Jepson (1943), whose treatment was on the whole fairly satisfactory. Nevertheless several changes seemed in order, both as regards the classification of certain critical species and the nomenclature of the sections. The genus as treated here falls rather naturally into six sections.

Below the level of section, species groups have been distinguished in *Gilia* sections *Gilia, Saltugilia and Arachnion* (by Grant and Grant 1950–1956); and subsections in *Phlox* (by Wherry 1955).

MINOR SYSTEMATICS

The task of straightening out the species and subspecies of the Polemoniaceae is still in its initial stages. The following paragraphs provide a guide to systematic treatments available at this level together with some critical observations concerning certain genera. A very tentative list of valid species and synonyms is also given at the end of Chapter 5.

It should be noted in consulting the lists of species that their compiler adopts a conservative species concept. Where geographical replacement and intergradation can be demonstrated, I am in favor of collecting the various populations together into one polytypic species (for a discussion cf. Grant 1957). Of course in many parts of the family the state of taxonomic exploration is not yet advanced to the point where a modern polytypic species concept can be applied with any degree of accuracy.

1. *Cobaea*. The treatments by Hemsley (1880), Brand (1907) and House (1908) have been superseded by Standley's revision of 1914. None of the species are known well in nature and many are known only from the type collection.

2. *Cantua*. The only taxonomic account is by Brand (1907). Ecological and distributional data are given by Weberbauer (1945).

3. *Huthia*. See Brand (1908, 1913b).

4, 5. *Bonplandia, Loeselia*. See Brand (1907) and Standley's *Trees and Shrubs of Mexico*.

6. *Polemonium*. Despite the fact that *Polemonium* has been treated by numerous authors (Greene 1887, Brand 1907, Wherry 1942, 1944a, 1945a, Davidson 1950, Vassiljev 1953a, 1953b, and Klokov 1955), the genus is still poorly understood and requires much further study. Wherry recognized 44 species, which is too many since some condensation of allopatric taxa into polytypic species seems justified, as has been done by Davidson. On the other hand Davidson's reduction of the genus to 19 species may go too far in the

opposite direction. Davidson placed some well-marked species like *P. brandegei* and *P. confertum* in synonymy and omitted one species, *P. flavum*, entirely. The number of species in my tentative list is 23.

Vassiljev (1953a, 1953b) has recognized 14 species of *Polemonium* in the U.S.S.R. and Klokov (1955) 13 species in Eurasia. It is probable on the basis of their studies that the genus *Polemonium* is more complex in eastern Europe and Asia than we have hitherto supposed. Several of the entities treated as species by the Russian authors, however, are evidently equivalent to subspecies in a polytypic species concept. Klokov created two binomials for different morphological types derived from the same hybrid combination of *Polemonium caeruleum* × *P. boreale*. The total number of species of *Polemonium* in Eurasia is almost certainly less than the 13 or 14 estimated by Vassiljev and Klokov, but may well be more than the four indicated in the present conservative treatment.

7. *Allophyllum*. The genus has been treated recently by Grant and Grant (1955).

8, 9. *Collomia, Gymnosteris*. See Mason (1951) and Munz (1957) for taxonomic descriptions and Wherry (1944) for a discussion of the entities. Additional notes on *Collomia* are given by Payson (1924). Concerning *Collomia* in South America see Reiche (1910) and Borsini (1942).

10. *Phlox*. The treatments of the western Phloxes by Nelson (1899) and St. John (1936), of the Texas annual Phloxes by Whitehouse (1945), of the eastern Phloxes by Wherry in a series of studies from 1929 to 1945, and of the genus as a whole by Brand (1907), has now been superseded by Wherry's monograph (1955). As noted by Wherry himself much work remains to be done in clarifying the minor taxa and their interrelationships in *Phlox*.

11. *Microsteris*. Greene (1898) recognized 7 species but Kinch (1956) (see also Munz 1957) in a recent study found that the various forms could be grouped into one species composed of two subspecies. With regard to this group in South America see Reiche (1910) and Borsini (1942).

12. *Gilia*. There is no single taxonomic account of the whole genus at present. The sections *Gilia, Saltugilia* and *Arachnion* have been revised by Grant and Grant (1950–1956). The other Pacific coast species are treated by Munz (1957). For the section *Giliastrum* and the main bulk of *Giliandra* one must consult various floras of the Rocky Mountain region and Southwest. The South American species are discussed by Reiche (1910), Borsini (1942) and Weberbauer (1945). Valuable comments on the California species are found in Jepson's *Flora* (1943).

13. *Ipomopsis*. The taxonomy of this genus is at present in an unsatisfactory state. Incomplete accounts exist for different species groups by Wherry (1946), Constance and Rollins (1936), and Kearney and Peebles (1943). The

South American species is discussed by Reiche (1910) and Borsini (1942). A general conspectus of the genus is given by Grant (1956).

14. *Eriastrum.* The latest taxonomic treatments are by Mason (1945, 1951) and Munz (1957). Supplementary material is found in Jepson (1943), Craig (1934) and Macbride (1917).

15. *Langloisia.* The standard treatments are by Mason (1951) and Munz (1957). Additional notes are found in Greene (1896) and Jepson (1943).

16. *Navarretia.* The taxonomy of this genus has been covered by Mason (1951) and Munz (1957). Additional notes on the California species are given by Greene (1887) and Jepson (1943), on the South American species by Reiche (1910) and Borsini (1942).

17. *Leptodactylon.* This genus has been covered by Jepson (1943), Wherry (1945), Mason (1951) and Munz (1957). The *L. pungens* complex needs much further study.

18. *Linanthus.* Good treatments are given by Mason (1951) and Munz (1957). For the Chilean species see Reiche (1910), for a compariosn of it with a related California species see Howell (1938). Additional notes on the genus are given by Jepson (1943), Wherry (1945b) and Greene (1892).

CHAPTER IV

SYSTEM OF CLASSIFICATION

Polemoniaceae

Perennial or annual herbs or less commonly shrubs, vines, or small trees. Leaves alternate or opposite, simple or pinnately or palmately compound. Flowers axillary or terminal, solitary or in small cymose clusters or dense heads, perfect. Calyx pentamerous, synsepalous, regular or sometimes zygomorphic, wholly herbaceous or with herbaceous lobes and membranous sinuses. Corolla pentamerous, sympetalous, regular or rarely bilabiate, aestivation contorted. Stamens 5, epipetalous, alternate with corolla lobes, the point of insertion varying, anthers 2-celled, dehiscing longitudinally. Ovary superior, inserted on a basal disk, 3-celled, rarely 2- or 4-celled, style single, stigma 3-lobed or rarely 2- or 4-lobed. Fruit a capsule, mostly loculicidal, with 1 to numerous seeds. Seeds mostly with copious endosperm and straight or slightly curved embryo, the seed coat frequently becoming mucilaginous when wet.

America, extending with 5 species into Eurasia. Eighteen genera and about 316 species.

SYNOPSIS OF THE TRIBES

I. Tropical

Woody plants (and some herbs in *Loeselia*). Leaves broad (except *Huthia*), alternate (or opposite in *Loeselia grandiflora*), the upper ones not reduced (except *Loeselia grandiflora*). Calyx herbaceous (except *Loeselia*), not rupturing in age (except *Loeselia*). Corolla large to medium-sized. Stamens inserted in base or on tube of corolla, two whorls of stamen traces present (unknown in *Huthia, Bonplandia, Loeselia*). Cells of capsule with numerous seeds each. Seeds winged, usually flat, with little or no endosperm, the coat not becoming mucilaginous when wet. Embryo white or yellowish, the cotyledons large and fleshy. Chromosomes minute (unknown in *Huthia* and *Bonplandia*).

A. Tribe Cobaeeae Baill.

Lianes. Leaves pinnately compound, the lateral leaflets broad, the terminal one modified as a tendril. Flowers solitary on long peduncles. Calyx regular, herbaceous, the lobes only slightly united at base. Corolla campanulate with oval or caudate lobes. Stamens inserted in base of corolla. Pollen yellow. Capsule much longer than calyx, dehiscing septicidally. Seeds large, flat, broadly winged.

Tropical America. One genus, *Cobaea*.

B. Tribe Cantueae Peter

Small trees and shrubs. Leaves simple or pinnately divided. Flowers axillary or clustered in terminal corymbs. Calyx regular, herbaceous, the tube well developed. Corolla tubular. Stamens inserted in base of corolla or on tube. Pollen blue or yellow. Capsule much longer than calyx, dehiscing loculicidally. Seeds large, flat, broadly winged.

Andes. Two genera, *Cantua* and *Huthia*.

C. Tribe Bonplandieae Baill.

Shrubs, subshrubs and herbs. Leaves simple, broad, margins serrate to pinnately divided. Flowers axillary, solitary or in pairs or clustered, or in panicles. Calyx slightly zygomorphic, herbaceous or with membranous sinuses. Corolla more or less zygomorphic. Stamens inserted on corolla tube. Pollen purple or perhaps yellow. Capsule about equalling to much shorter than calyx, dehiscing loculicidally. Seeds small, broadly to very narrowly winged.

Tropical America. Two genera, *Bonplandia* and *Loeselia*.

II. Temperate

Mainly herbaceous plants, sometimes woody. Leaves broad or narrow, alternate or opposite, the upper ones reduced or not reduced. Calyx wholly herbaceous or with membranous sinuses. Corolla medium-sized to small. Stamens inserted on wall of corolla tube or above, one whorl of stamen traces being present. Cells of capsule with 1 to numerous seeds each. Seeds not winged, plump, containing endosperm, the coat becoming mucilaginous when wet or unchanged. Embryo yellowish to bright green, the cotyledons smaller, foliaceous or fleshy. Chromosomes large to small but not minute.

Table 2. The Tropical Genera of the Polemoniaceae compared

Genus	Habit	Leaves	Inflorescence	Calyx	Corolla	Stamens	Capsule	Seed
Cobaea	lianes	pinnately compound with a terminal tendril	flowers solitary on long peduncles, axillary	regular, herbaceous, sepals divided to base	regular, campanulate	inserted in base of corolla	much longer than calyx, dehiscence septicidal	flat broadly winged
Cantua	shrubs and small trees	simple, broad	flowers in terminal corymbs	regular, herbaceous, sepals fused nearly to apex	regular, tubular	inserted in base of corolla	much longer than calyx, dehiscence loculicidal	flat, broadly winged
Huthia	shrubs	pinnately divided, long and narrow	flowers axillary or terminal in small clusters	regular, herbaceous, sepals fused nearly to apex	regular, tubular	inserted in corolla tube	much shorter than calyx, dehiscence loculicidal	flat, broadly winged
Bonplandia	shrubs	broad with serrate margins, or pinnately divided	flowers axillary in pairs	slightly zygomorphic, herbaceous, sepals fused nearly to apex	bilabiate, salverform	inserted in corolla tube	much shorter than calyx, dehiscence loculicidal	plump, very narrowly winged
Loeselia	shrubs and herbs	simple, broad	flowers axillary or in panicles	slightly zygomorphic, membranous in the sinuses, sepals fused nearly to apex	slightly zygomorphic, salverform or tubular	inserted in corolla tube	about equalling calyx, often enveloped by bracts, dehiscence loculicidal	flat or plump, broadly or narrowly winged

D. Tribe Polemonieae Baill.

Subshrubs, perennial herbs, or less commonly annuals. Herbage dark green or grayish. Leaves pinnately compound or simple, the dissection usually not complex, often narrow or linear; upper cauline leaves not much reduced. Calyx usually herbaceous (except *Phlox, Microsteris, Allophyllum*). Corolla regular. Stamens inserted in corolla tube or throat, insertion regular or commonly very irregular. Pollen color usually constant within a genus, usually yellow (blue in *Allophyllum* and at least one species of *Collomia*).

Cold temperate North America with extensions into Eurasia, temperate South America, and warmer regions of North America. Six genera, *Polemonium Allophyllum, Collomia, Gymnosteris, Phlox, Microsteris*.

E. Tribe Gilieae Reichb.

Subshrubs, perennial herbs, or more commonly annuals. Herbage often bright green but also dark green or grayish. Leaves simple but often dissected in a complex manner, or entire and linear; upper cauline leaves frequently strongly reduced. Calyx with membranous sinuses which rupture in age. Corolla regular to bilabiate. Stamens inserted on corolla tube,. throat or in sinuses of lobes, insertion mostly regular but where irregular not strongly so. Pollen color frequently varying within a genus, blue, yellow or white.

Arid southwest of North America with extensions to the northern Rocky Mountain region, to the Pacific and Atlantic coasts, and into the deserts and mountains of temperate South America. Seven genera, *Gilia, Ipomopsis, Eriastrum, Langloisia, Navarretia, Leptodactylon, Linanthus*.

A. Tribe Cobaeeae

1. Cobaea Cav.

Lianes. Leaves alternate, pinnately compound, the lateral leaflets broad, the terminal one modified as a tendril, or the first seedling leaves frequently with an unmodified terminal leaflet. Flowers axillary, solitary on long peduncles. Calyx regular, herbaceous, sepals divided to the base, not rupturing in age. Corolla large, regular, campanulate, with oval or caudate lobes, violet, yellow, brownish-purple or green. Stamens inserted in base of corolla. Pollen yellow, greenish or brownish. Capsule much longer than calyx, septicidal. Seeds large, flat, broadly winged.

Montane rain forests from Mexico to Venezuela and in the Andes from Colombia to northern Chile. The approximately 19 species are mostly known from one or a few highly localized collections.

Cobaea scandens is in cultivation in Europe, North America, Brazil, Java, etc. (Bonstedt 1932, Bailey 1933), and occasionally escapes from cultivation. Concerning *Cobaea* in Europe, Brazil, and Java see respectively Hegi (1927), Brand (1907), and Backer (1951).

(i) Section Cobaea   Fig. 19

Flowers erect. Corolla lobes ovate or orbicular, without caudate tips. Stamens equalling or longer than corolla lobes.

Mexico to the Andes. In Mexico, Central America, Colombia, and Venezuela: *Cobaea biaurita, lasseri, lutea, minor, pachysepala, pringlei, scandens, skutchii, stipularis, trianaei, triflora, viorna.* In Peru and northern Chile: *C. campanulata.*

(ii) Section Aschersoniophila Brand   Fig. 20

Flowers pendant. Corolla lobes ovate at the base with long caudate tips. Stamens about equalling corolla lobes.

Costa Rica. *Cobaea aschersoniana.*

(iii) Section Rosenbergia (Oersted) Brand   Fig. 21

Flowers pendant. Corolla lobes linear and caudate. Stamens usually longer, sometimes shorter than corolla lobes.

Costa Rica to Venezuela. *Cobaea aequatoriensis, gracilis, hookeriana, panamensis, penduliflora.*

B. Tribe Cantueae

2. Cantua Juss. *Fig. 22*

Shrubs and small trees. Leaves alternate, simple, broad. Flowers in terminal corymbs. Calyx regular, herbaceous, tubular, not rupturing in age. Corolla large, regular, tubular, red or white. Stamens inserted in base of corolla. Pollen blue or yellow. Capsule much longer than calyx, loculicidal. Seeds large, flat, broadly winged.

Andes from Ecuador to Bolivia. Six species, *Cantua bicolor, buxifolia, candelilla, ochroleuca, pyrifolia, quercifolia.*

Cantua buxifolia is cultivated as an ornamental in Peru, the United States and Europe (Bonstedt 1932, Bailey 1933). This species, which was a sacred plant of the Incas used to adorn their temples, is scarcely known in a wild state in Peru (Weberbauer 1945). The leaves of *Cantua pyrifolia* when pressed in water yield a soapy solution which was used by the Peruvian indians for washing clothes (Brand 1907).

Fig. 19. Cobaea scandens. Calyx in bud on left. (from Peter.)

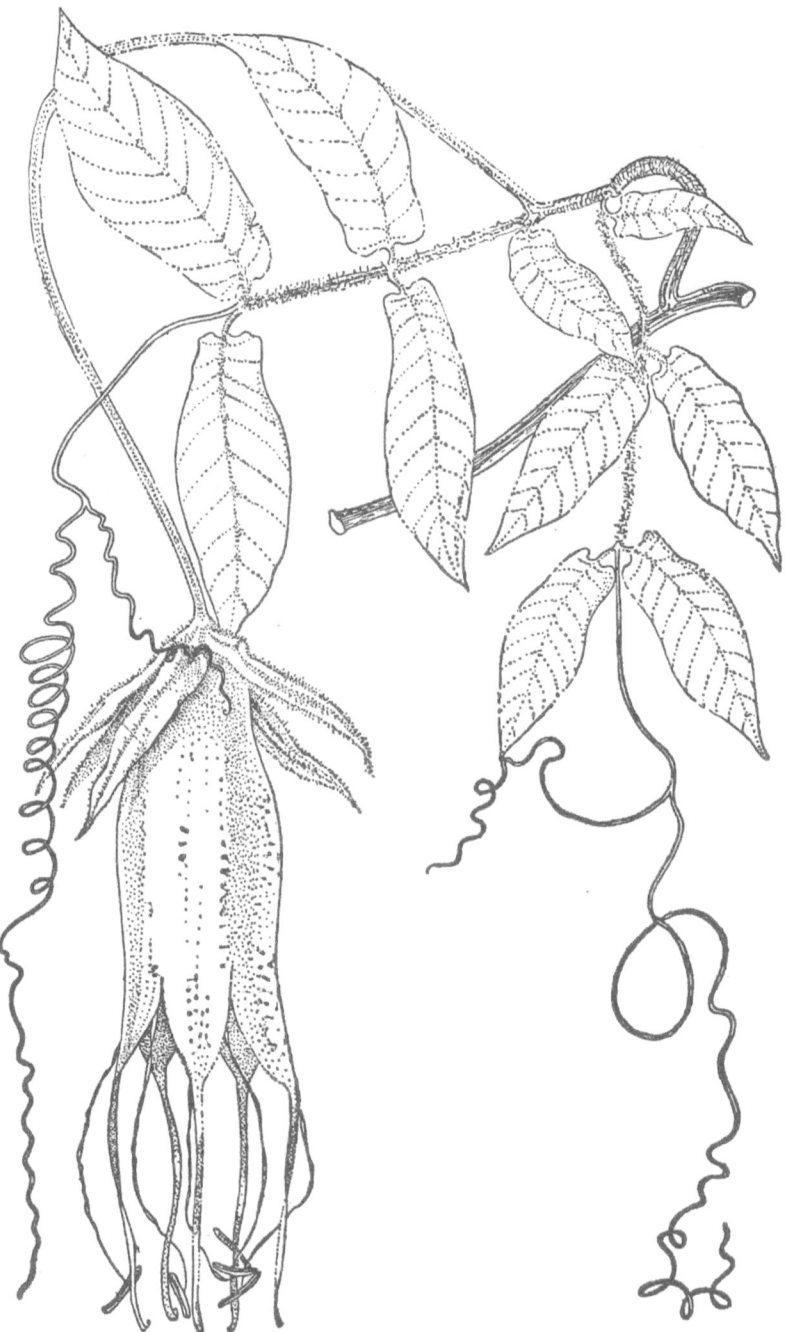

Fig. 20. Cobaea aschersoniana (redrawn from Brand.)

Fig. 21. Cobaea penduliflora.
Seed on right. (from Brand; flower parts enlarged, habit reduced × .5.)

Fig. 22. Cantua quercifolia.
Fruit, seed and embryo of C. pyrifolia on left. (from Brand; embryo × 5.)

3. Huthia Brand   *Fig. 23*

Shrubs. Leaves alternate, pinnately divided, broadly linear. Flowers axillary or terminal in small clusters. Calyx regular, herbaceous, tubular, not rupturing in age. Corolla large to medium-sized, regular, tubular, violet.

Fig. 23. Huthia coerulea. (redrawn from Weberbauer.)

Stamens inserted on corolla tube. Pollen blue. Capsule much shorter than calyx, loculicidal. Seeds large, flat, broadly winged.

Peru. Two species, *Huthia coerulea, longiflora.*

C. Tribe Bonplandieae

4. Bonplandia Cav.   *Fig. 24*

Shrubs. Leaves alternate, simple, broad, pinnately divided or nearly entire with serrate margins. Flowers axillary in pairs. Calyx slightly zygomorphic, herbaceous, tubular, not rupturing in age. Corolla medium-sized, bilabiate, salverform, violet. Stamens inserted on corolla tube. Capsule much shorter than calyx, loculicidal. Seeds small, plump, very narrowly winged.

Mexico. One or two species, *Bonplandia geminiflora, linearis*.

5. Loeselia L.

Shrubs, woody-based perennials and annual herbs. Leaves simple, broad, with serrate margins. Flowers axillary or in panicles. Calyx slightly zygomorphic, tubular, differentiated into herbaceous lobes and membranous sinuses, rupturing in age. Corolla medium-sized, nearly regular to somewhat zygomorphic, salverform or funnelform. Stamens inserted on corolla tube. Pollen purple or perhaps yellow. Capsule about equalling calyx, often enveloped by bracts, loculicidal. Seeds small, flat or plump, broadly or narrowly winged (unknown in section *Glumiselia*).

Southern Arizona to Colombia and Venezuela. Approximately 9 species.

A tea made from the leaves of *Loeselia mexicana* is much used in Mexico to alleviate fevers. The herbage of this plant was formerly crushed in water to make a soapy solution (Brand 1907, Standley 1924).

(*i*) Section Loeselia   *Fig. 25*

Low shrubs, woody-based perennial herbs, and annuals. Leaves alternate, medium-small, the upper ones not much reduced. Flowers axillary, the bracts green and leaf-like. Corolla funnelform or tubular–funnelform with oval lobes, red or violet.

Southern Arizona to Colombia and Venezuela. *Loeselia amplectens, ciliata, coerulea, glandulosa, involucrata, mexicana, pumila, scariosa*.

(*ii*) Section Glumiselia Grant

Shrubs up to 8 feet tall. Leaves opposite, large, the upper ones much reduced. Flowers in an open branched panicle, with several dark purplish bracts closely appressed around the base of each flower so as to insheathe the calyx and corolla tube. Corolla salverform with a very narrow tube and long linear clawed lobes, white.

West Central Mexico. *Loeselia grandiflora*.

Fig. 24. Bonplandia geminiflora. Calyx at lower left. (from Brand; calyx × 2.)

Fig. 25. Loeselia pumila.
Calyx at lower left, flower with bract at lower right. (from Brand; calyx and flower × 4.)

D. Tribe Polemonieae

6. Polemonium L.

Rhizomatous or cespitose perennial herbs, and one annual. Herbage often villous or glutinous and with a skunk-like odor. Leaves alternate, pinnately compound, the leaflets entire or divided. Flowers in terminal or axillary cymes, solitary to capitate. Calyx herbaceous, accrescent, not rupturing in age. Corolla showy to small, rotate, campanulate, funnelform or tubular, blue, purple, yellow or white. Stamens regularly inserted on corolla tube, equal in length, exserted or included. Pollen yellow or white. Cells of capsule with 1–10 seeds each. Seeds medium-sized, fusiform, angular, shiny, dark, unchanged when wet (except in one species).

North America, Eurasia, Patagonia and Chile. Approximately 23 species.

Polemonium caeruleum is widely cultivated in European gardens and frequently escapes from cultivation (Hegi 1927). Several other species are also in use in Europe and North America as border plants or rock garden subjects (Bonstedt 1932, Bailey 1933, Wherry 1944). *Polemonium caeruleum* and *P. boreale* were used as antisyphiletics by European peasants and Siberians (Brand 1907, Hegi 1927).

(i) Section Polemonium   Fig. 26

Perennial herbs. Leaflets coplanar. Corolla mostly campanulate but sometimes rotate or tubular, showy or rarely small. Seeds more or less unchanged when wet.

North America to Eurasia. In North America: *Polemonium boreale, caeruleum, californicum, carneum, delicatum, flavum, foliosissimum, glabrum, grandiflorum, mexicanum, pauciflorum, pectinatum, pulcherrimum, reptans*. In Eurasia: *P. boreale, caeruleum, humile, pulchellum*.

(ii) Section Melliosma Peter ex Brand   Fig. 27

Perennial herbs. Leaflets verticillate. Corolla mostly funnelform, medium-sized. Seeds unchanged when wet.

Mountains of western North America. *Polemonium brandegei, chartaceum, confertum, elegans, eximium, viscosum*.

(iii) Section Polemoniastrum Peter ex Brand   Fig. 28

Annuals. Corolla shorter than the calyx. Seeds mucilaginous when wet.

Great Basin of North America and Chile and Patagonia. *Polemonium micranthum*.

Table 3. The Genera of the

Genus	Duration	Leaves	Calyx
Polemonium	mostly perennials and 1 annual	alternate, pinnately compound	herbaceous, not rupturing in age
Allophyllum	annuals	alternate, pinnate or simple	differentiated into herbaceous lobes and membranous sinuses, rupturing in age
Collomia	perennials and annuals	alternate, mostly simple and entire, sometimes lobed or dissected	herbaceous with carinate sinuses, not rupturing in age
Gymnosteris	diminutive annuals	leafless	scarious but lobes and sinuses not differentiated, not rupturing in age
Phlox	perennials and a few annuals	mostly opposite, simple and entire	differentiated into herbaceous lobes and membranous sinuses, the latter often carinate, rupturing in age, regular
Microsteris	annual	opposite, simple and entire	differentiated into herbaceous lobes and membranous sinuses, the latter not carinate rupturing in age, slightly zygomorphic

Tribe Polemonieae compared

Corolla	Stamens	Capsule	Seeds	Basic ch. no.
rotate, campanulate or funnelform	insertion regular, equal in length	cells 1–10–seeded	medium-sized, shiny, dark, fusiform, angular, unchanged when wet (except Polemoniastrum)	$x = 9$
funnelform	insertion regular or irregular, equal or unequal in length	cells 1–3–seeded	large dark spheroidal, mucilaginous when wet	$x = 9,8$
funnelform to salverform	insertion regular or irregular, equal or unequal in length	cells mostly 1-seeded	large, dark, lenticular, mucilaginous when wet (except Collomiastrum)	$x = 8$
salverform	insertion regular, sessile, equal	cells several-seeded	quite fine, light colored, slightly mucilaginous	$x = 6$
salverform, inner base of tube hairy	insertion irregular, unequal in length	cells mostly 1–seeded	large, lenticular unchanged when wet	$x = 7$
small and inconspicuous, inner base of tube not or only slightly hairy	insertion irregular	cells 1–seeded	large, lenticular,, mucilaginous when wet	$x = 7$

7. Allophyllum (Nutt.) Grant   *Fig. 29*

Annual herbs. Herbage often villous or glutinous and with a skunk-like odor. Leaves alternate, pinnately or bipinnately divided or lobed or simple and entire, the upper cauline leaves digitately lobed. Flowers in small terminal loose to congested glomerules. Calyx differentiated into herbaceous lobes and membranous sinuses, accrescent, but rupturing in age. Corolla medium to small but never showy, funnelform, blue-violet, red-violet or white. Stamens regularly or irregularly inserted on corolla tube, equal or unequal in length, exserted or included. Pollen blue. Fruiting pedicels usually much elongated. Cells of capsule with 1–3 seeds each. Seeds large, ovoidal, black, mucilaginous when wet.

Pacific slope and southwestern region of North America. Five species, *Allophyllum divaricatum, gilioides, glutinosum, integrifolium, violaceum.*

Fig. 26. Polemonium humile. Stamen and ovary of P. flavum. (from Peter.)

Fig. 27. Polemonium viscosum. (from Davidson;
habit × .75; flower × 1.5.)

8. Collomia Nutt.

Annual or rhizomatous perennial herbs. Herbage sometimes glutinous and with a skunk-like odor. Leaves alternate, mostly simple and entire, linear to lanceolate, or sometimes pinnately lobed. Flowers in loose or congested terminal cymes, or rarely solitary and axillary. Calyx herbaceous, sometimes becoming chartaceous with age, sinuses carinate, the calyx enlarging with the fruit and not rupturing in age. Corolla showy to small and inconspicuous, funnelform to salverform, blue, purple, red, salmon-yellow or white. Stamens regularly or irregularly inserted on corolla throat or tube, equal or unequal in length, mostly included, sometimes exserted. Pollen yellow or blue. Cells of capsule with 1–2 seeds each. Seeds large, lenticular, dark, mostly mucilaginous when wet, sometimes unchanged.

North and South America. Fourteen species.

Collomia grandiflora has become naturalized locally in central Europe (Hegi 1927) and *C. linearis* in Russia and Australia (Vassilev 1953, Ewart 1930).

Fig. 28. Polemonium micranthum.
Flower and fruit on lower left. (from Brand; flower and fruit × 2; pistil and corolla × 3.)

Fig. 29. Allophyllum divaricatum. (from Brand; habit × .67; flower × 3.)

(*i*) *Section Collomiastrum Brand*   *Fig. 30*

Perennials with slender rootstocks. Leaves mostly broad. Corolla funnel-form. Cells of capsule single-seeded. Seeds more or less unchanged when wet.

Mountains of western North America. *Collomia debilis, larseni, mazama, rawsoniana.*

Fig. 30. Collomia debilis. (from Abrams; habit × 1.)

(ii) Section Courtoisia (Reichb.) Wherry   *Fig. 31*

Annuals. Leaves broad. Corolla funnelform to salverform. Cells of capsule several-seeded. Seeds becoming mucilaginous when wet.

Pacific slope of North America. *Collomia diversifolia, heterophylla.*

Fig. 31. Collomia heterophylla. (from Abrams; habit × 1.)

(iii) Section Collomia   *Fig. 32*

Annuals. Leaves lanceolate to linear. Corolla funnelform to salverform. Cells of capsule mostly single-seeded. Seeds becoming mucilaginous when wet.

North and South America. In western North America: *Collomia grandiflora, linearis, macrocalyx, tenella, tinctoria, tracyi*. In northeastern North America: *C. linearis*. In Chile and Patagonia: *C. biflora, cavanillesii*.

Fig. 32. Collomia tinctoria.
Seed, embryo and fruiting calyx at lower right. (from Brand; flower × 3; seed × 7; calys × 5.)

9. Gymnosteris Greene *Fig. 33*

Diminutive annuals. Foliage consisting of a pair of persistent connate cotyledons and a whorl of 4–5 larger lanceolate-ovate entire bracts united at the base; true foliage leaves absent or sometimes present in the form of minute linear segments in the axils of the bracts. Flowers borne in a small terminal head subtended by the involucre-like whorl of bracts. Calyx urceolate, vescicular, the tissue of the sinuses not differentiated from that of the lobes, not rupturing in age. Corolla small and inconspicuous, salverform, yellow or white. Stamens regularly inserted on corolla throat, sessile. Pollen yellow. Cells of capsule with several to many seeds each. Seeds minute, light brown, becoming slightly mucilaginous when wet.

Western North America. Two species, *Gymnosteris nudicaulis, parvula.*

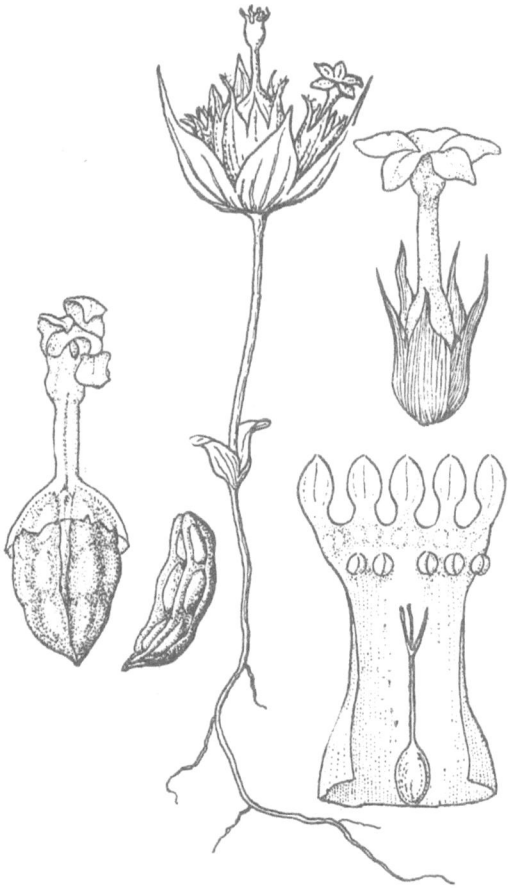

Fig. 33. Gymnosteris parvula. (from Abrams; habit × 5.)

10. Phlox L.

Perennial herbs, subshrubs, and a few annuals. Leaves opposite, or the uppermost ones sometimes alternate, sessile, simple and entire, obovate to linear or subulate. Flowers mostly grouped in terminal cymes or panicles or sometimes solitary. Calyx regular, differentiated into herbaceous lobes and membranous sinuses, sinuses often carinate, rupturing in age. Corolla showy, salverform, the tube often bent, inner base of tube hairy, purple, violet, pink or whitish. Stamens irregularly inserted on corolla tube, unequal in length, mostly included or rarely slightly exserted. Pollen orange, yellow or white. Cells of capsule with 1–2 or rarely 5 seeds each. Seeds large to small, lenticular, unchanged when wet.

North America and Siberia. Sixty-one species.

Many of the species and their hybrids are in cultivation as garden plants. Good horticultural accounts are given by Bonstedt (1932), Bailey (1933), Clay (1937), Ingwersen (1948), Symons-Jeune (1953) and Wherry (1935, 1946, 1955). *Phlox drummondii* has escaped from cultivation in many parts of the world (Wherry 1935).

(i) Section Phlox   Fig. 34

Tall perennial herbs or sometimes low herbs, rarely with a somewhat woody base. Leaves deciduous, large to medium-sized. Flowers numerous or sometimes few in cymes or panicles. Stamens long, the upper ones subequalling to slightly exceeding the corolla tube. Style exceeding the calyx and often subequalling the corolla tube, stigma lobes quite short. Seeds large.

North America. In eastern North America: *Phlox amplifolia, bifida, buckleyi, carolina, glaberrima, maculata, ovata, paniculata, pulchra, stolonifera, subulata.* In western North America: *P. adsurgens, amabilis, caryophylla, cluteana, dolichantha, idahonis, longifolia, stansburyi, viscida.*

(ii) Section Divaricatae Peter   Fig. 35

Perennial herbs or woody-based perennials or a few annuals. Leaves deciduous, obovate to linear. Flowers numerous or sometimes few in cymes or panicles. Stamens short, included. Style shorter than calyx, stigma lobes equalling or much exceeding the style. Seeds large.

North America. In eastern North America: *Phlox amoena, divaricata, floridana, nivalis, oklahomensis, pilosa.* In western North America: *P. colubrina, grahami, mesoleuca, nana, speciosa, tenuifolia, triovulata.* In Texas: *P. cuspidata, drummondii, glabriflora, pilosa, roemeriana, triovulata,* and several species listed for eastern North America. In Mexico: *P. mesoleuca, mexicana, nana, tenuifolia, triovulata.*

Table 4. The Sections of the Genus Phlox compared

Section	Duration	Habit	Leaves	Inflorescence	Stamens	Style and Stigma	Seeds
Phlox	perennial	herbs, or rarely woody at base	deciduous, large to medium-sized	flowers mostly numerous in cymes or panicles	long, equalling corolla or exserted	style long, stigma short	large
Divaricatae	perennial or annual	herbs, or woody at base	deciduous, large to medium-sized	flowers mostly numerous in cymes or panicles	short, included	style short, stigma long	large
Occidentales	perennial	subshrubs, often cespitose	evergreen, medium-sized to needle-like	flowers solitary or few, terminal on branches	short, included	style short, stigma short	medium-sized or small

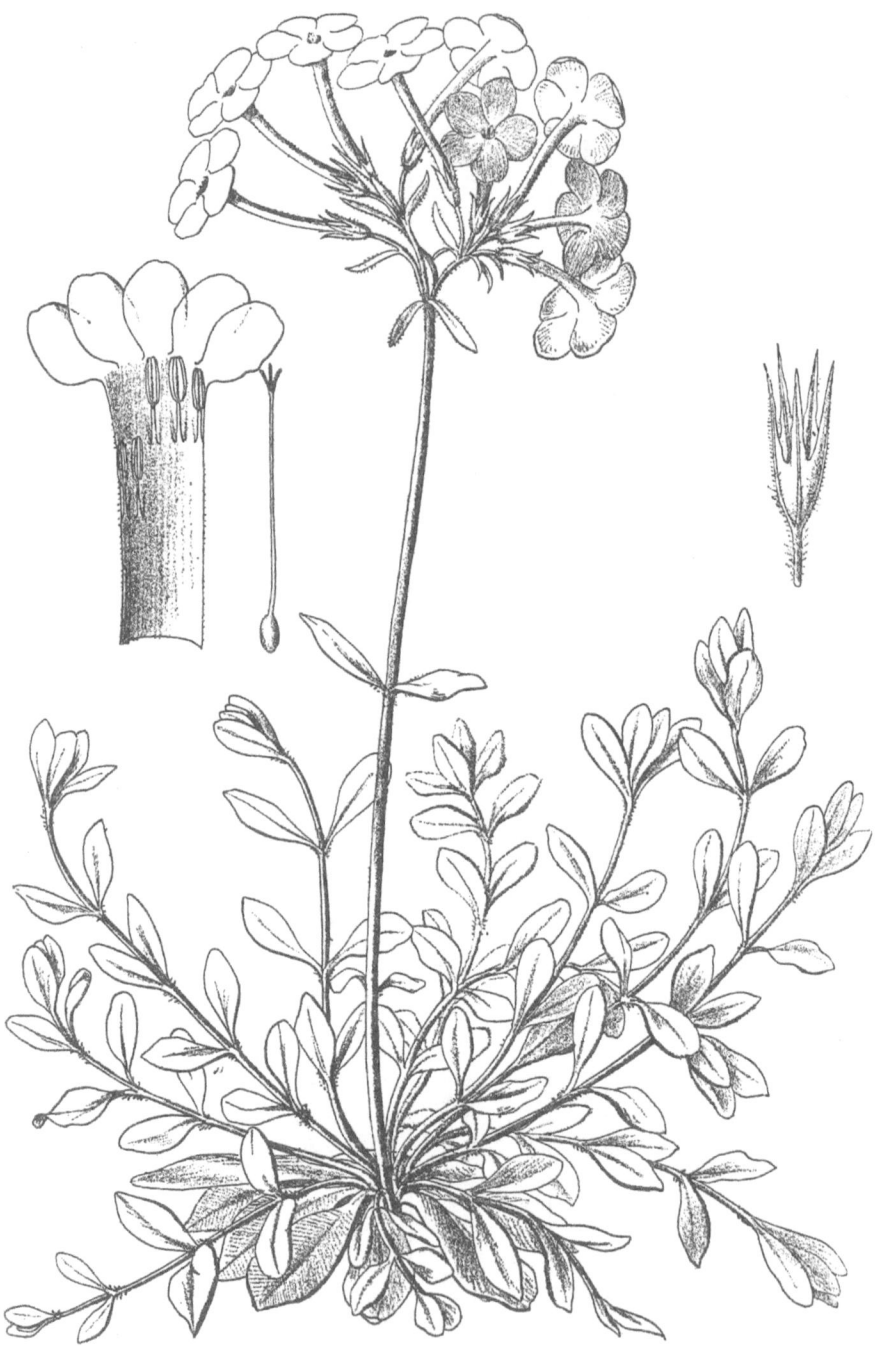

Fig. 34. Phlox stolonifera. (from Brand; corolla and pistil × 2; calyx × 3.)

Fig. 35. Phlox nana. (from Brand; calyx × 2.)

(iii) Section Occidentales Gray   *Fig. 36*

Cespitose or cushion-like subshrubs. Leaves evergreen, medium-sized to needle-like, hard and stiffish. Flowers solitary or in small groups at ends of branches. Stamens short, included. Style shorter than or sometimes subequalling calyx, stigma lobes quite short. Seeds medium-sized or small.

North America and Asia. In western North America: *Phlox aculeata, albomarginata, alyssifolia, andicola, austromontana, bryoides, caespitosa, covillei, diffusa, douglasii, gladiformis, griseola, hoodii, jonesii, kelseyi, missoulensis, mollis, multiflora, peckii, variabilis.* In Alaska: *P. borealis, richardsonii.* In central and northeastern Asia: *P. sibirica.*

11. Microsteris Greene   *Fig. 37*

Small annual herbs. Leaves opposite, except the upper ones alternate, simple and entire, obovate to linear. Flowers grouped in pairs in upper axils. Calyx slightly zygomorphic, differentiated into herbaceous lobes and membranous sinuses, sinuses not carinate, rupturing in age. Corolla small and inconspicuous, salverform, inner base of tube not or only slightly hairy, violet to white. Stamens irregularly inserted on corolla tube, included. Pollen yellowish. or whitish. Cells of capsule with 1 seed each. Seeds large, lenticular, mucilaginous when wet.

North and South America. One polytypic species, *Microsteris gracilis.*

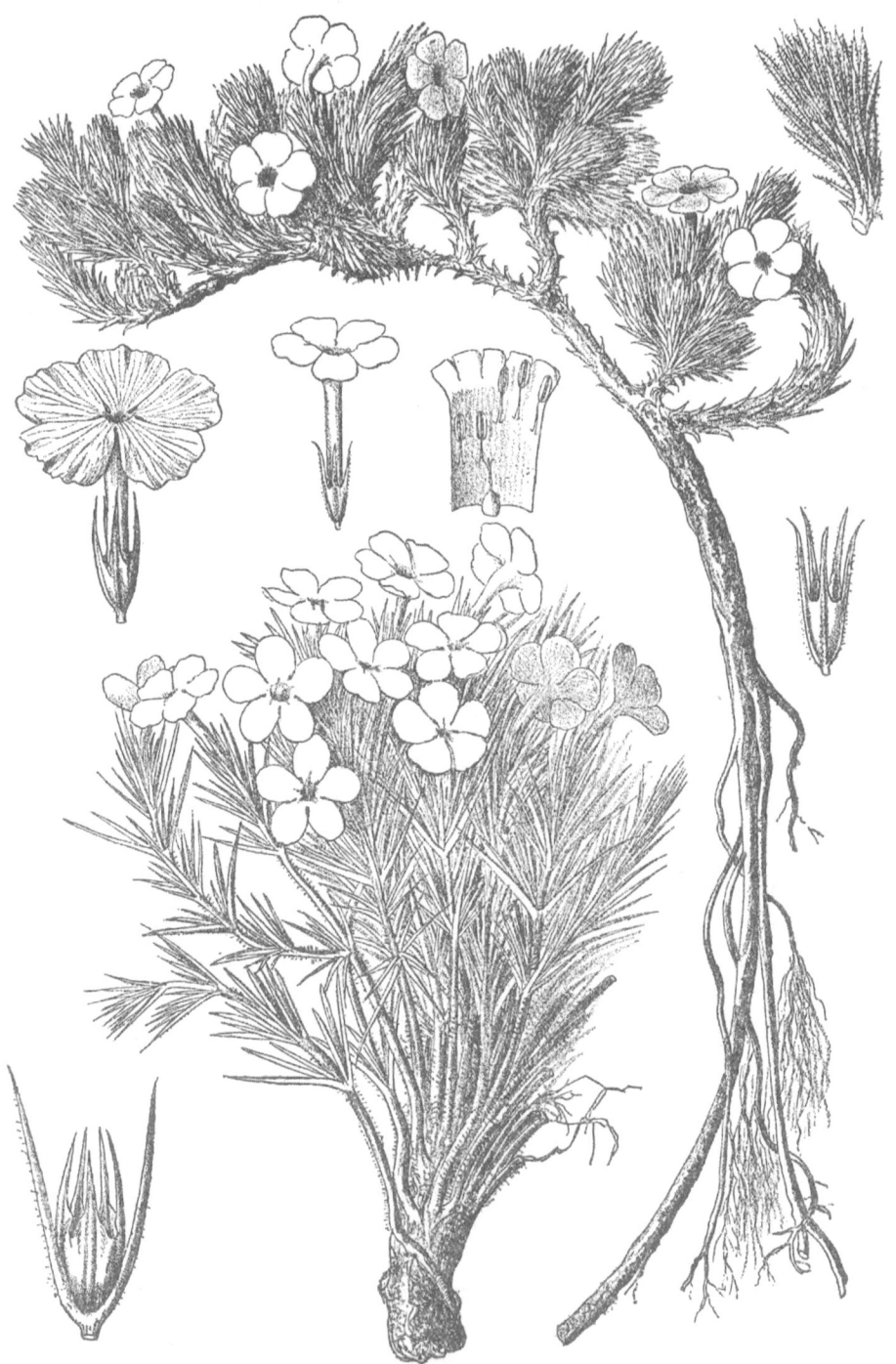

Fig. 36. Phlox caespitosa (above) and P. austromontana (below).
(from Brand; flower, corolla and leaves of P. caespitosa × 2; calyx of P. caespitosa × 3; flower of
P. austromontana × 2; calyx of P. austromontana × 3.)

Fig. 37. Microsteris gracilis. Part of calyx on left. (from Brand; habit × .67; flower × 3.)

E. Tribe Gilieae

12. Gilia R. & P.

Annual herbs and some perennial and biennial herbs. Herbage usually bright green. Leaves alternate, pinnately dissected or sometimes entire and linear, mostly well developed below and becoming strongly reduced above. Flowers in loose or glomerate cymes or in dense heads or sometimes solitary, blooming in spring. Calyx regular. Corolla showy to small and inconspicuous, regular, mostly funnelform or sometimes subsalverform, violet, pink, white or rarely yellow. Stamens mostly regularly, sometimes slightly irregularly inserted on corolla tube, throat or sinuses of lobes, equal or mostly slightly unequal in length, exserted or included, anthers oval and erect. Pollen blue or white or less commonly yellow. Cells of capsule with 1 to many seeds each. Seeds small or minute, ovoidal, brown, mucilaginous when wet.

North and South America. Fifty-six species.

Several species, particularly *G. capitata*, *G. achilleaefolia* and *G. tricolor*, are grown as garden oranementals (Bonstedt 1932, Bailey 1933) and occasionally escape from cultivation (Grant and Grant 1956).

(i) Section Giliastrum Brand   Figs. 38,43

Perennials, sometimes woody-based, or annuals, sometimes small and delicate. Pubescence of herbage stipitate-glandular or consisting of multicellular hairs. Leaves broad to linear, if pinnatifid not more than once dissected, soft to hard and stiffish but not leathery, bright green to grayish-green; middle cauline leaves much like the lower ones. Corolla rotate to campanulate, concolored (or rarely bicolored with streaks in throat), lobes blue, pale violet, pink or yellow. Stamens inserted in base of corolla or midway in throat, exserted or included. Pollen yellow or blue.

Southwestern plains and deserts of North America and deserts and mountains of temperate South America. In North America: *Gilia campanulata*, *effusa*, *filiformis*, *incisa*, *inyoensis*, *latifolia*, *palmeri*, *purpusii*, *rigidula*, *ripleyi*, *uncialis*. In South America: *G. foetida* and probably the *G. glutinosa* group (*G. chachanensis*, *cobijanensis*, *glabrata*, *glutinosa*, *ramosissima*).

(ii) Section Giliandra Gray   Fig. 39

Perennials, biennials or annuals. Pubescence of herbage stipitate-glandular, scurfy, or consisting of multicellular hairs. Leaves once pinnatifid with a broad rachis and short broad lobes, dark green and leathery; middle cauline leaves reduced and usually entire. Corolla funnelform, concolored, lobes pink to white. Stamens inserted on corolla tube or throat or less commonly in

Table 5. The Genera of the Tribe Gilieae compared

Genus	General Features	Distribution of Leaves	Phyllotaxy and Leaf form	Calyx	Corolla	Stamens	Basic Chromosome Number
Gilia	spring-blooming annuals, and some perennial and biennial herbs	mostly scapose with strongly reduced upper cauline leaves	alternate, pinnately dissected, some-times linear	regular	regular, mostly funnelform	insertion mostly regular or rarely irregular, anthers oval, erect	x = 9, chromosomes large
Ipomopsis	summer-blooming perennials, desert annuals, & one shrub	plants leafy throughout	alternate, pinnatifid to linear	regular	mostly regular, sometimes slightly zygomorphic, salverform with broad tube	insertion mostly regular, or irregular, anthers oval, erect	x = 7, chromosomes large
Eriastrum	summer-blooming annuals and one shrub	lower stems frequently leafless, the upper stems very leafy	alternate, pinnately dissected or linear	zygomorphic	mostly regular sometimes zygomorphic, salverform, funnelform	insertion regular, anthers sagittate appearing oval in reduced forms, versatile	x = 7
Langloisia	spring and summer-blooming desert annuals	as in Eriastrum	alternate, pinnately toothed, bristle tippèd	regular, spiny	regular or strongly bilabiate, funnel-form	insertion regular, anthers oval, erect or versatile	x = 7
Navarretia	summer annuals of open fields, meadows and vernal pools	as in Eriastrum	alternate, pinnately lobed, prickly	zygomorphic spiny	regular, mostly funnelform	insertion regular or irregular, anthers oval, erect or versatile	x = 9
Lepto-dactylon	shrubs	plants leafy throughout	alternate or opposite, palmately divided	regular	regular, salverform with narrow tube	insertion regular, anthers oval, erect	x = 9, chromosomes small
Linanthus	spring-blooming annuals and one perennial herb	plants leafy throughout	opposite, palmately divided	regular	regular, mostly salverform with narrow tube	insertion regular anthers oval, erect	x = 9 chromosomes small

sinuses of lobes, mostly very long exserted, sometimes included. Pollen blue.

Rocky Mountain region and western deserts of North America. *Gilia hutchinsifolia, leptomeria, mcvickerae, micromeria, pentstemonoides, pinnatifida, stenothyrsa, subnuda.*

(iii) Section Gilia. Leafy-stemmed Gilias   Fig. 40

Annuals of large to medium size. Pubescence of herbage consisting of stipitate-glandular hairs or multicellular hairs or both. Leaves once, twice or thrice pinnately dissected, bright green and soft-textured; middle cauline leaves reduced but usually dissected; upper cauline leaves relatively well developed. Corolla funnelform, concolored or tricolored with solid spots in throat, lobes blue-violet. Stamens inserted in sinuses of corolla lobes, short exserted or included. Pollen blue.

Pacific North America and temperate South America. In North America: *Gilia achilleaefolia, angelensis, capitata, clivorum, millefoliata, nevinii, tricolor.* In South America: *G. laciniata, valdiviensis.*

(iv) Section Arachnion Grant. Cobwebby Gilias   Fig. 41

Annuals of medium to small size. Pubescence of herbage arachnoid-wooly. Leaves once, twice or thrice pinnately dissected, bright green and soft-textured; middle cauline leaves reduced but usually dissected; upper cauline leaves much reduced. Corolla funnelform to subsalverform, concolored or more commonly bi- or tricolored with solid spots or rings in throat, lobes blue-violet, pink or white. Stamens inserted in sinuses of corolla lobes, short exserted or included. Pollen blue.

Deserts and bordering mountain ranges in North and South America. In North America: *Gilia aliquanta, brecciarum, cana, diegensis, inconspicua, interior*[1], *latiflora, leptantha, mexicana, minor, modocensis, ochroleuca, ophthalmoides, sinuata, tenuiflora, transmontana, tweedyi.* In South America: *G. crassifolia.*

(v) Section Saltugilia Grant. Woodland Gilias   Figs. 42,44

Annuals of large to small size. Pubescence of herbage consisting of stipitate-glandular hairs or multicellular hairs or both or geniculate hairs. Leaves once, twice or thrice pinnately dissected, or reduced to linear lobes, bright green and soft-textured; middle and upper cauline leaves reduced or well developed. Corolla funnelform, concolored or bi- or tricolored with dots in throat, corolla lobes mostly pink, sometimes violet. Stamens regularly or irregularly inserted on corolla throat or sinuses of lobes, short exserted or included. Pollen blue.

---

[1] **Gilia interior** (Mason & A. Grant) A. Grant, **comb. nov.** (*Gilia tenuiflora* ssp. *interior* Mason & A. Grant, Madroño 9: 217, 1948. *Gilia inconspicua* ssp. *interior* A. & V. Grant, El Aliso 3: 253, 1956).

Fig. 38. Gilia rigidula. Calyx at lower left. (from Brand; calyx × 4.)

Pine woods and bordering communities in Pacific North America, interior deserts, rare in Rocky Mountains. *Gilia australis, capillaris, caruifolia, leptalea, scopulorum, splendens, stellata, tenerrima.*

Table 6. The Sections of the Genus Gilia compared

	Giliastrum	Giliandra	Gilia	Arachnion	Saltugilia
Duration	perennial, annual	perennial, biennial, annual	annual	annual	annual
Pubescence	glandular, or multi-cellular hairs	glandular, scurfy, or multi-cellulair hairs	glandular, or multi-cellular hairs	arachnoid woolly	glandular, or multi-cellular hairs, or geniculate hairs
Lower Leaves	broad or linear and once pinnate	once pinnate with a broad rachis, leathery	1–3-pinnate	1–3-pinnate	1–3-pinnate or linear
Upper Leaves	like the lower ones	reduced and usually entire	reduced but not strongly so	strongly reduced	reduced or well developed
Corolla Form	rotate or campanulate	funnelform	funnelform	funnelform or sub-salverform	funnelform
Corolla Color	concolored, or rarely bicolored with streaks	concolored	concolored or tri-colored with spots	mostly bi- or tri-colored with spots and rings, sometimes concolored	bi- or tricolored with dots, or concolored
Stamen Insertion	regular, in base of corolla or midway in throat	regular, on tube, throat or sinuses	regular, in sinuses	regular, in sinuses	regular or irregular, on throat or sinuses
Stamen Length	exserted or included	mostly long exserted	short-exserted or included	short-exserted or included	short-exserted or included
Pollen	yellow or blue	blue	blue	blue	blue

Fig. 39. Gilia pinnatifida. (from Brand;
habit × .5; flower and pistil × 3.)

Fig. 40. Gilia capitata.
(original, by Mrs. K. Goss.)

Fig. 41. Gilia sinuata. (from Abrams; habit × 1.)

Fig. 42. Gilia leptalea. (from Abrams; habit × 1.)

Fig. 43. Gilia latifolia. (from Abrams; habit reduced slightly.)

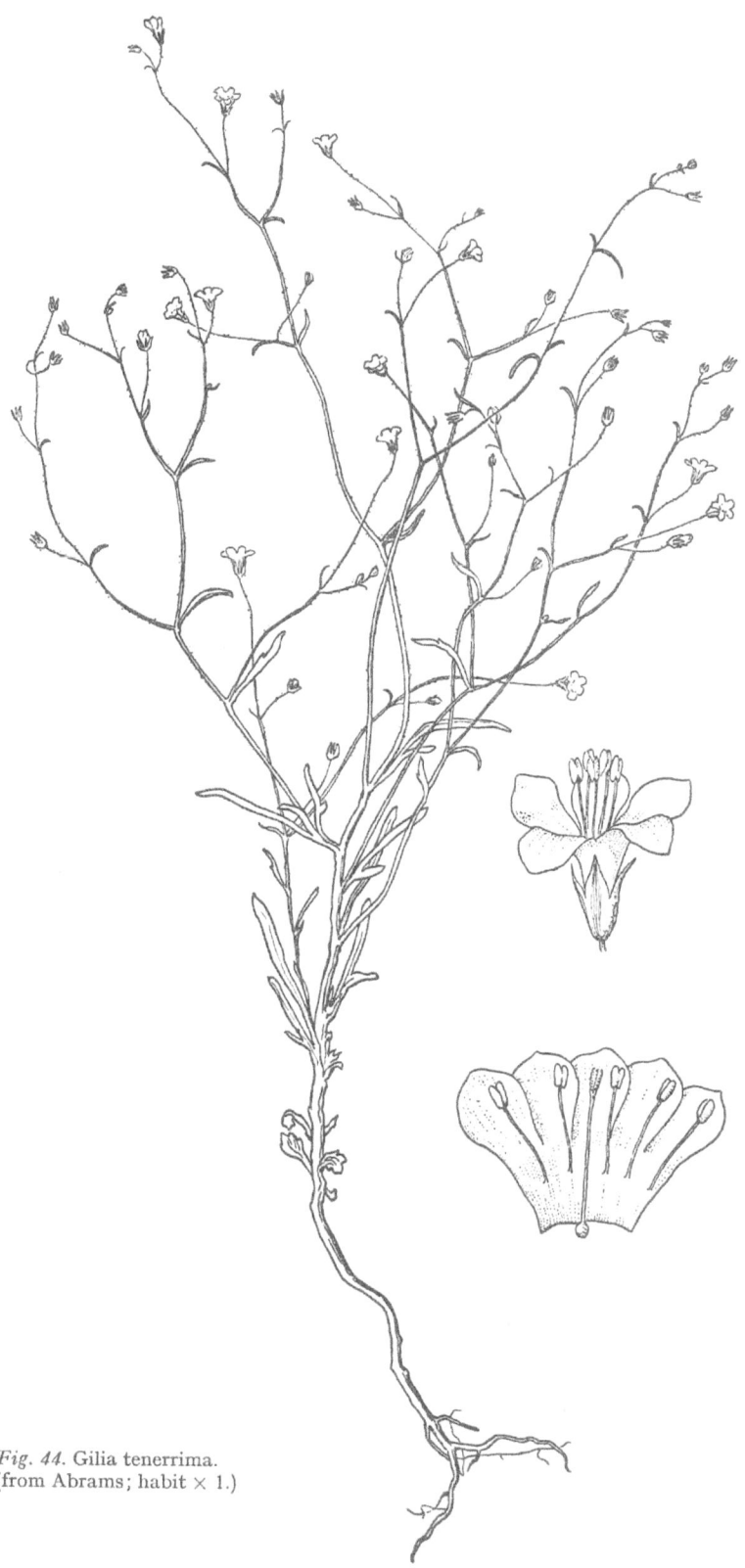

Fig. 44. Gilia tenerrima.
(from Abrams; habit × 1.)

13. Ipomopsis Michx.

Perennial or biennial herbs, a few annuals, and one shrub. Herbage often grayish-green or blue-green. Leaves alternate, pinnatifid to entire and linear, the tips of the segments with horny mucros. Stems clothed with well developed leaves throughout. Flowers in loose racemes or dense heads, the individual flowers subtended by bracts, blooming in summer. Calyx regular. Corolla large and showy to small and inconspicuous, regular or sometimes slightly bilabiate, tubular or salverform with a broad tube, red,. pink, violet, cream or white. Stamens mostly regularly, sometimes irregularly inserted on corolla tube or sinuses of lobes, equal or unequal in length, exserted or included, anthers oval and erect. Pollen blue or yellow. Cells of capsule with 1 to several seeds each. Seeds mostly long, slender, bent, corrugated, whitish, rarely ovoidal and brown, mostly mucilaginous when wet, sometimes unchanged.

North and South America. Twenty-four species.

Several species are very attractive and have been brought into cultivation to some extent as garden ornamentals (Bonstedt 1932, Bailey 1933). *Ipomopsis macombii* is used as a fish-poison by the indians in northern Mexico.

(i) Section Phloganthea (Gray) Grant   Fig. 45

Spreading perennials with numerous slender stems from base, lacking a conspicuous basal rosette of leaves in mature plants, or low rounded shrubs. Corolla slightly zygomorphic, tube long to medium long, corolla violet, red-violet or white. Stamens long exserted. Pollen blue.

Southwestern North America and Mexico. *Ipomopsis gloriosa, havardi, multiflora, pinnata, polyantha, tenuifolia.*

(ii) Section Ipomopsis   Fig. 46

Perennials, biennials and two annuals. Habit erect and fairly tall with one to several stout stems from base, a basal rosette present. Corolla regular, showy, the tube long, corolla red, pink, yellow or white. Stamens exserted or included. Pollen blue or yellow.

Rocky Mountain region to Florida and the Pacific coast. In western North America: *Ipomopsis aggregata, laxiflora, longiflora, macombii, tenuituba, thurberi.* In southeastern North America: *I. rubra.*

(iii) Section *Microgilia* (*Benth.*) *Grant* — Figs. 47,48

Annuals and perennials. Habit low with simple stems or branching from base, a basal rosette present or absent. Corolla regular, the tube quite short, individual flowers inconspicuous but flowering heads moderately conspicuous, corolla white or violet. Stamens short, slightly exserted or included. Pollen blue.

Rocky Mountain region to the Pacific slope and Mexico, and in Argentina and Chile. In North America: *Ipomopsis congesta, depressa, frutescens, gunnisonii, minutiflora, polycladon, pumila, roseata, sonorae, spicata.* In South America: *I. gossypifera.*

14. Eriastrum Woot. & Standl.   *Fig. 49*

Annual herbs and one perennial subshrub. Herbage often more or less densely wooly. Leaves alternate, pinnately dissected or entire and linear, lower stems frequently naked and upper stems very leafy. Flowers mostly in bracteate heads, rarely solitary, blooming in summer. Calyx zygomorphic, the lobes pungent. Corolla medium-sized to small, mostly regular or sometimes bilabiate, funnelform, blue, or sometimes yellow or white. Stamens regularly inserted on corolla throat or sinuses of lobes, exserted or included, anthers sagittate, appearing oval in reduced forms, versatile. Pollen blue or yellow. Cells of capsule with 1 to several seeds each. Seeds ovoidal, brown, mucilaginous when wet.

Western North America. Fourteen species, *Eriastrum abramsii, brandegeae, densifolium, diffusum, eremicum, filifolium, hooveri, luteum, pluriflorum, sapphirinum, sparsiflorum, tracyi, virgatum, wilcoxii.*

Eriastrum densifolium is occasionally cultivated as an ornamental (Bailey 1933).

15. Langloisia Greene   *Fig. 50*

Low annual herbs. Herbage grayish-green. Leaves alternate, pinnately toothed, bristle-tipped, lower stems frequently naked and upper stems very leafy. Flowers in small terminal bracteate heads, blooming in spring and summer. Calyx regular, the lobes spine-tipped. Corolla medium-sized to small, regular or strongly bilabiate, funnelform, pink, violet, yellowish or whitish, often with purple specks. Stamens regularly inserted on corolla throat, equal or unequal in length, mostly exserted, anthers oval, erect or versatile. Pollen blue or yellow. Cells of capsule with 2–9 seeds each. Seeds small, ovoidal, brown, mucilaginous when wet.

North American deserts. Five species, *Langloisia lanata, matthewsii, punctata, schottii, setosissima.*

Table 7. The Sections of the Genus Ipomopsis compared

Section	Duration	Habit	Corolla form	Corolla color	Stamens
Phloganthea	perennial	spreading herbs with numerous stems from base, one low shrub, no basal rosette at maturity	slightly zygomorphic, large or medium-sized	violet, red-violet, or white	long exserted, pollen blue
Ipomopsis	perennial, biennial, annual	erect and fairly tall herbs, basal rosette present	regular, large	red, pink, yellow, or white	exserted or included, pollen blue or yellow
Microgilia	perennial, annual	low herbs, basal rosette present or absent	regular, small	white or violet	included or short-exserted, pollen blue

Fig. 45. Ipomopsis tenuifolia. (from Brand; habit × 1; flower × 2.)

Fig. 46. Ipomopsis aggregata. (original by S. Tillett; habit × .4; flower × .8; seed × 9.)

Fig. 47. Ipomopsis spicata. (from Brand; habit × .5; flower × 3.)

Fig. 48. Ipomopsis congesta. (from Abrams; habit × 1; seed × 10.)

Fig. 49. Eriastrum densifolium (left) and E. filifolium (right).
(from Brand; habit × .67; flowers and bracts enlarged.)

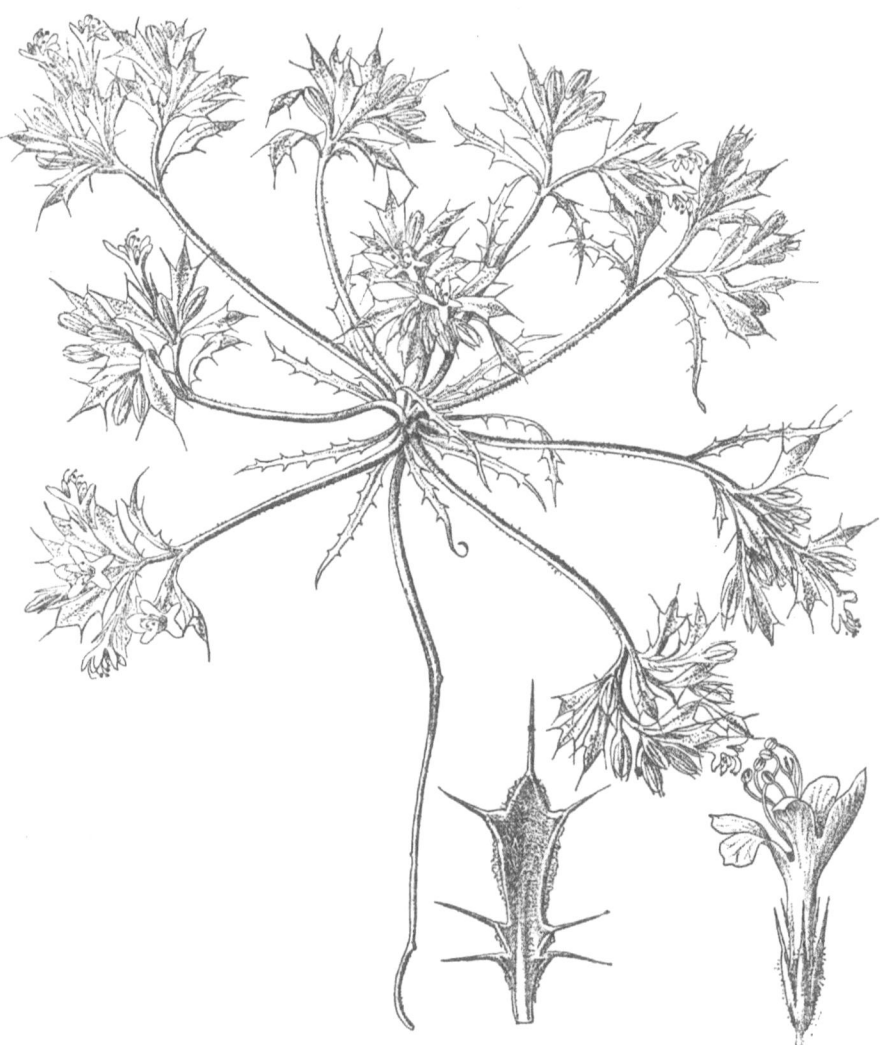

Fig. 50. Langloisia matthewsii. (from Brand; leaf and flower × 3.)

16. Navarretia R. & P.

Annual herbs. Herbage frequently glutinous, sometimes glabrous. Leaves alternate, pinnately lobed or sometimes entire and linear, spine-tipped, lower stems frequently naked and upper stems very leafy. Flowers in heads subtended by spiny bracts, blooming in summer. Calyx zygomorphic, the lobes spine-tipped. Corolla small, regular, mostly funnelform or sometimes salverform, purple, blue, violet, pink, yellow or white. Stamens regularly or irregularly inserted on corolla throat or sinuses of lobes, equal or slightly unequal in length, exserted or included, anthers oval, erect or versatile. Pollen blue or whitish. Cells of capsule with 1 to many seeds each. Seeds small, ovoidal, often minutely pitted, brown, mucilaginous when wet.

North and South America. Thirty species.

Navarretia squarrosa is naturalized as a weed in South America, Australia and probably elsewhere in the world (Ridley 1930, Ewart 1930).

Fig. 51. Navarretia atractyloides. (from Abrams; habit × 1.)

Table 8. *The Sections of the Genus Navarretia compared*

	Aegochloa	Masonia	Mitracarpium	Navarretia
Habit	tall, central leader strong	medium-tall to small, central leader short, lower branches stronger	medium-tall to small, central leader strong	medium-tall to small, central leader strong or short
Pubescence	glandular	glandular to minutely glandular	minutely glandular or glabrous	puberulent or glabrous
Leaves	once pinnate with a broad or narrow rachis and rigidly spiny	once pinnate with linear rachis, more lax and less rigidly spiny than in Aegochloa	bipinnate with a linear rachis, soft-textured	once or twice pinnate with a narrow rachis, soft-textured
Bracts	base broad and heavy, rigidly spiny	as in Aegochloa	base narrow, not rigidly spiny	narrow, long, much exceeding heads, soft-textured
Calyx	sinuses without a tuft of hairs	as in Aegochloa	as in Aegochloa	sinuses bearing a tuft of hairs
Corolla	brightly colored, lobes 3–veined	brightly colored, lobes 3–veined	brightly colored, lobes usually 3–veined, sometimes 1–veined	pale colored, lobes usually 1–veined, rarely 3–veined
Stamens	inserted on corolla throat	inserted on corolla throat	inserted on corolla throat	inserted on throat or sinuses
Stigma	3–lobed	3–lobed	3– or 2–lobed	2–lobed
Capsule	dehiscence loculicidal from top or base, 3– or rarely 2–celled, pericarp leathery	dehiscence loculicidal from base or rarely from top, 3–celled, pericarp leathery or papery	dehiscence circumscissile or rarely indehiscent, 1–celled, pericarp papery or membranous	indehiscent, 1–celled, pericarp membranous
Seeds	many to few in each locule	many to few in each locule	solitary or few in locules	solitary or few in locules

(i) *Section Aegochloa (Benth.) Grant*   *Fig. 51*

Plants tall with a strong central leader. Herbage glandular, odorous. Middle cauline leaves once pinnate with a broad rachis or rarely linear, rigidly spiny; bracts with a broad heavy base, rigidly spiny. Calyx sinuses not bearing a tuft of hairs. Corolla brightly colored, blue, purple or yellow; lobes 3-veined. Stamens inserted on corolla throat. Stigma 3-cleft. Capsule loculicidally dehiscent, from top downwards or base upwards, 3-celled or rarely 2-celled, pericarp leathery. Cells of capsule with few to many seeds each.

Open fields and meadows, western North America. *Navarretia atractyloides, filicaulis, hamata, heterodoxa, mellita, squarrosa.*

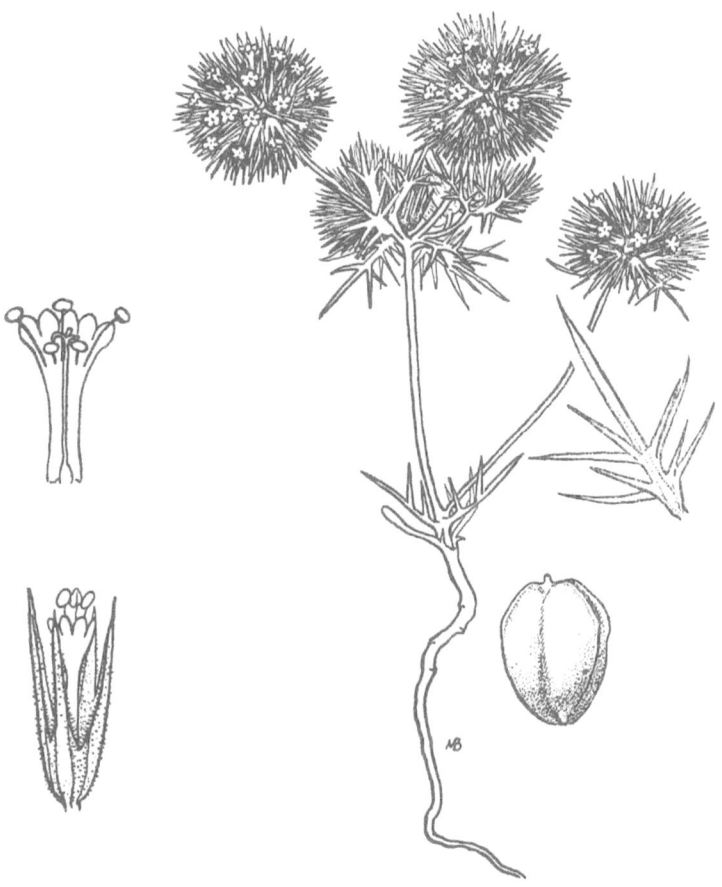

Fig. 52. Navarretia breweri. (from Abrams; habit × 1.)

Fig. 53. Navarretia eriocedhala. (from Abrams; habit × 1.)

(ii) Section Masonia Grant   Fig. 52

Plants medium-tall to minute, central leader short, branches from beneath becoming stronger. Herbage glandular to minutely glandular-puberulent. Middle cauline leaves once pinnate, rachis linear, the lobes fewer and more lax, less rigidly spiny than in Section *Aegochloa*; bracts with a broad heavy base and rigidly spiny. Calyx sinuses not bearing a tuft of hairs. Corolla brightly colored, purple, yellow or pink; lobes 3-veined. Stamens inserted on corolla throat. Stigma 3-cleft. Capsule loculicidally dehiscent, from base upwards or rarely from top downwards, 3-celled, pericarp leathery or papery. Cells of capsule with few to many seeds each.

Open fields and meadows, western North America. *Navarretia breweri, divaricata, peninsularis, prolifera, viscidula.*

(iii) Section Mitracarpium Brand   Fig. 53

Plants medium-tall to minute, central leader usually strong. Herbage minutely or not at all glandular. Middle cauline leaves bipinnate with a narrow linear rachis, soft-textured, not rigidly spiny; bracts with a narrow base, not rigidly spiny. Calyx sinuses not bearing a tuft of hairs. Corolla brightly colored, blue, purple or yellow; lobes usually 3-veined, sometimes with 1 branched vein. Stamens inserted on corolla throat. Stigma 3- or 2-cleft. Capsule mostly dehiscent and circumscissile or sometimes indehiscent, 1-celled, pericarp papery to membranous. Seeds few to solitary in capsule.

Open fields and meadows, western North America. *Navarretia cotulaefolia, eriocephala, heterandra, intertexta, jepsonii, mitracarpa, nigellaeformis, propinqua, pubescens, setiloba, tagetina.*

(iv) Section Navarretia   Fig. 54

Plants medium-tall to minute, central leader strong or short and branched from below. Herbage puberulent or practically glabrous. Middle cauline leaves once or twice pinnate, rachis narrow and linear, soft, not rigidly spiny; bracts narrow, soft, long, much exceeding the heads. Calyx sinuses bearing a tuft of hairs. Corolla pale blue or white; lobes linear, usually with 1, or rarely with 3 veins. Stamens inserted in corolla sinuses or throat. Stigma 2-cleft, the lobes minute[2]. Capsule indehiscent, 1-celled, pericarp membranous. Seeds few to solitary in capsule.

[2] Crampton's (1954) report of entire stigmas in this section is not documented and seems doubtful; at least the observation cannot be confirmed from examination of material in the Claremont herbarium. One is reminded of a previous discovery of entire stigmas in *Gilia* sect. *Gilia* by Heller (1906), which subsequent examination showed to be the result of selecting immature stages (Grant 1950).

Vernal pools, North and South America. In western North America: *Navarretia bakeri, leucocephala, minima, pauciflora, pleiantha, prostrata, subuligera*. In Argentina and Chile: *N. involucrata*.

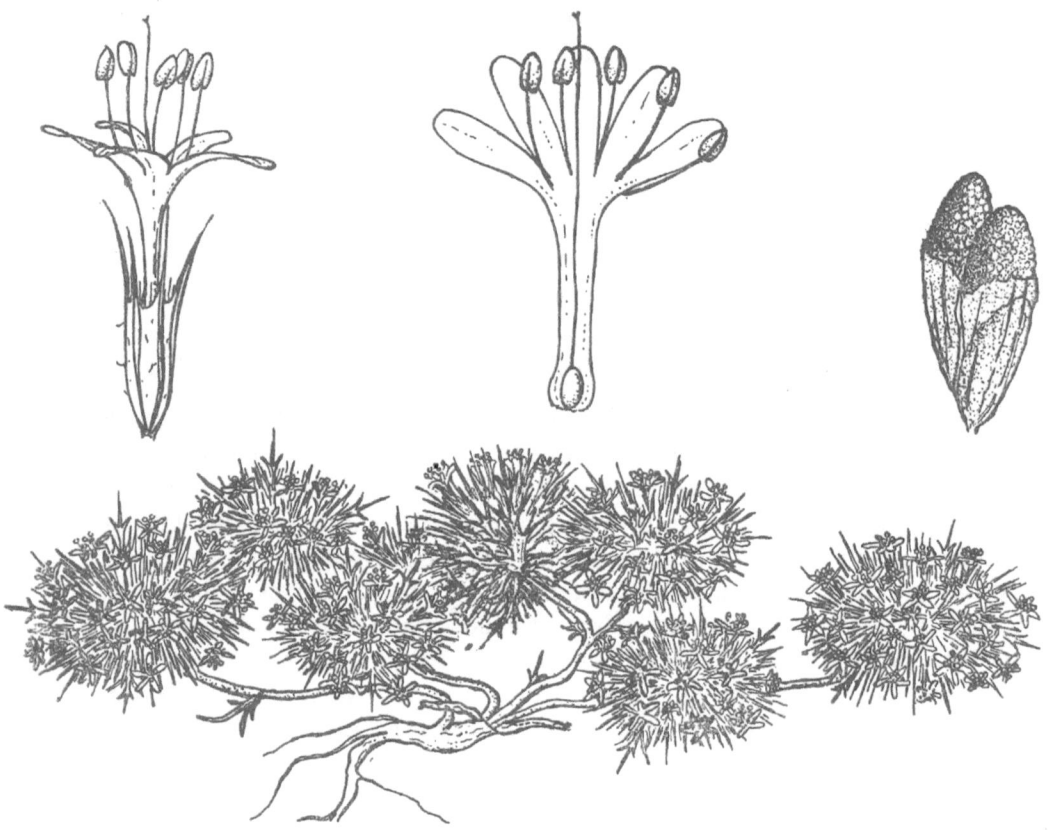

Fig. 54. Navarretia pleiantha. (from Abrams; habit × 2.)

17. Leptodactylon H. & A.   *Fig. 55*

Subshrubs. Leaves alternate or opposite, palmately divided, the lobes pungent, stems clothed with well developed leaves throughout. Flowers in terminal congested cymes or glomerules or rarely solitary and axillary. Calyx regular, the lobes pungent. Corolla often showy, regular, mostly salverform with a narrow tube, pink, light violet, white or cream. Stamens regularly inserted on corolla tube or throat, equal or subequal in length, included, anthers oval and erect. Pollen yellowish or whitish. Cells of capsule with several to many seeds each. Seeds small, ovoidal, brown.

Mountains of western North America. Six species, *Leptodactylon caespitosum, californicum, jaegeri, pungens, veatchii, watsonii*.

Fig. 55. Leptodactylon californicum. (from Brand; leaf × 2.)

18. Linanthus Benth.

Annuals and two perennial herbs. Herbage usually bright green. Leaves opposite, palmately divided (rarely pinnately divided or simple), soft-textured, stems clothed with well developed leaves throughout. Flowers in terminal loose cymes or glomerules or sometimes solitary and axillary, blooming in spring. Calyx regular. Corolla showy to inconspicuous, regular, salverform with a narrow tube or funnelform or campanulate, pink, violet, yellow or

Fig. 56. Linanthus nuttallii. (from Abrams; habit × 1.)

white. Stamens regularly inserted on corolla throat or sometimes on tube or in sinuses of lobes, equal or sometimes unequal in length, exserted or included, anthers oval and erect. Pollen orange, yellow or white. Cells of capsule with 1 to several seeds each. Seeds small, ovoidal, brown, mucilaginous or unchanged when wet.

North and South America. Thirty-seven species.

Fig. 57. Linanthus grandiflorus. (from Abrams; habit × 1.)

Many of the species are quite attractive; as regards their horticulture see Bonstedt (1932) and Bailey (1933). A decoction made from the herbage of *Linanthus ciliatus* was used by the North American indians as a blood purifier and remedy for coughs (Brand 1907).

(*i*) *Section Siphonella (Gray) Grant*   *Fig. 56*

Perennial herbs and one annual. Leaves and bracts mostly long with numerous lobes. Flowers sessile or pedicelled. Calyx herbaceous with a very narrow membranous portion in the sinuses, cleft to about middle. Corolla funnelform, tube stout, tube about equalling calyx. Stamens inserted on corolla throat.

Fig. 58. Linanthus acicularis. (from Abrams; habit × 1.)

Western North America, mostly in the mountains. *Linanthus laxus, melingii*[3] *nuttallii.*

(ii) *Section Pacificus (Jeps.) Grant*   Fig. 57

Annuals. Leaves and bracts long, with numerous lobes. Flowers sessile. Calyx differentiated into herbaceous lobes and membranous sinuses, cleft to about middle, margin of sinus membrane bordering lobes slightly concave or

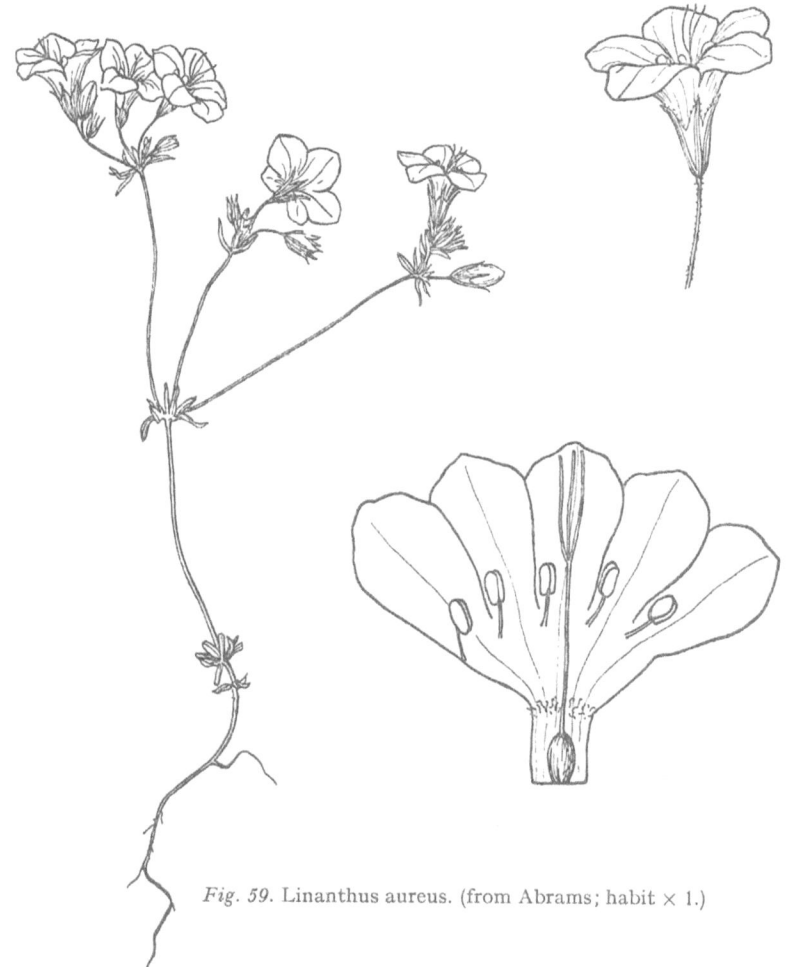

Fig. 59. Linanthus aureus. (from Abrams; habit × 1.)

[3] **Linanthus melingii** (Wiggins) V. Grant, **comb. nov.**
(*Leptodactylon Melingii* Wiggins, Contrib. Dudley Herb. 1: 173, t. 14, 16, 1933. *Linanthastrum melingii* Wherry, Amer. Midl. Nat. 34: 386, 1945).

Table 9. *The Sections of the Genus Linanthus compared*

	Siphonella	Pacificus	Leptosiphon	Dactylophyllum	Linanthus	Dianthoides
Duration	perennial, annual	annual	annual	annual	annual	low tufted annuals
Leaves and Bracts	mostly long, with numerous lobes	long, with numerous lobes	mostly long, with numerous lobes	quite short, with few to intermediate numbers of lobes	long, leaves sometimes simple	short, with few lobes, palmate or pinnate
Flowers	sessile or pedicelled	sessile	sessile	solitary on long peduncles, or sessile in one species	sessile or nearly so	sessile
Calyx	herbaceous with very narrow membranous portion in sinuses	differentiated into herbaceous lobes and membranous sinuses	as in Pacificus	as in Pacificus	as in Pacificus	as in Pacificus
Calyx Division	cleft to about middle	cleft to about middle	cleft to about middle	cleft to about middle	cleft to above middle	cleft to below middle or nearly to base
Calyx Sinus	membranous portion very narrow	membrane bordering lobes slightly concave or at least not rounded-convex	as in Pacificus	as in Pacificus	sinus membrane truncate or nearly so and quite broad	membrane bordering lobes rounded-convex
Corolla	funnelform, tube stout, tube about equalling calyx	as in Siphonella	salverform, tube very slender or filiform and long exserted from calyx	campanulate to funnelform, small to minute	funnelform, tube included in calyx, vespertine	campanulate to funnelform
Stamens	inserted on corolla throat	inserted on corolla throat	inserted on corolla throat	inserted on corolla throat or in sinuses of lobes	inserted on corolla tube	inserted on corolla throat

at least not rounded-convex. Corolla funnelform, tube stout, tube about equalling calyx. Stamens inserted on corolla throat.

Cismontane California. *Linanthus grandiflorus.*

(iii) Section Leptosiphon (Endl.) Grant   *Fig. 58*

Annuals. Leaves and bracts mostly long, with numerous lobes. Flowers sessile. Calyx as in section *Pacificus.* Corolla salverform, tube filiform, and long exserted from calyx. Stamens inserted on corolla throat.

Pacific slope of North America. *Linanthus aicicularis, androsaceus, bicolor, breviculus, ciliatus, montanus, nudatus, oblanceolatus, serrulatus.*

(iv) Section Dactylophyllum (Benth.) Grant   *Fig. 59*

Annuals. Leaves and bracts quite short with few to intermediate numbers of lobes. Flowers solitary on long peduncles, or sessile in one species. Calyx as in section *Pacificus.* Corolla campanulate to funnelform, small to minute. Stamens inserted on corolla throat or in sinuses of lobes.

Western North America and Chile. In western North America: *Linanthus ambiguus, aureus, bakeri, bolanderi, filipes, harknessii, lemmonii, liniflorus, pygmaeus, rattanii, septentrionalis.* In Chile: *L. pusillus.*

(v) Section Linanthus   *Fig. 60*

Annuals. Leaves and bracts long, sometimes simple. Flowers sessile or nearly so. Calyx differentiated into herbaceous lobes and membranous sinuses, cleft to above middle, sinus membrane truncate and quite broad. Corolla funnelform, tube included in calyx, vespertine. Stamens inserted on corolla tube.

Southwestern desert plains and arid coastal valleys. *Linanthus bigelovii, concinnus, dichotomus, jonesii.*

(vi) Section Dianthoides (Endl.) Grant   *Fig. 61*

Low tufted annuals. Leaves and bracts short, with few lobes, palmate or pinnate. Flowers sessile. Calyx differentiated into herbaceous lobes and

Fig. 60. Linanthus dichotomus. (from Abrams; habit × 1.)

membranous sinuses, cleft to below middle or nearly to base, margin of sinus membrane bordering lobes rounded-convex. Corolla campanulate to funnelform. Stamens inserted on corolla throat.

Southwestern desert plains and arid coastal valleys. *Linanthus arenicola, bellus, demissus, dianthiflorus, killippii, maculatus, orcuttii, parryae.*

Fig. 61. Linanthus parryae. (from Abrams; habit × 1.)

CHAPTER V

NOMENCLATURE

The present summary of nomenclature down to the level of sections and series is designed to accompany the system of classification set forth in the preceding chapter. It was felt that the taxonomy could stand out more clearly if the necessary but uninspiring task of clearing away the nomenclatural debris were reserved for a separate quarter. Chapter 5 was separated from chapter 4 in the interest of an unencumbered text.

A review of nomenclatural problems in the family Polemoniaceae as a whole has not been attempted for half a century. Brand's monograph of 1907 was written in an age before the method of nomenclatural types was generally accepted or even understood. One of the tasks of the present paper is to designate such types in many of the older infrageneric categories. It was also necessary to replace some of the sectional names proposed by recent authors by the older legitimate names. In such cases the rule of priority has been adhered to. In addition, two new sections (in *Navarretia* and *Loeselia*) and six new combinations for sectional names (in *Navarretia* and *Linanthus*) had to be made. Finally of course, many new taxa have been proposed since 1907 which needed to be cataloged.

Though the tables of synonyms given below have been arranged in accordance with a system of classification favored in this work, it is hoped that the usefulness of the present nomenclatural summary will not be confined to the adherents of any one system. The indication of the type species for each generic and subgeneric name should enable future workers motivated by different evidences and viewpoints to apply the correct names to their subdivisions with less bibliographical work and confusion than has befallen polemoniaceous taxonomists in the past.

Nomina nuda like *Brickellia* Raf. and *Myotoca* Griseb., and obvious misspellings like *Curtoisia* Endl. (for *Courtoisia* Reichb.), have not been included in the tables of synonyms. Pre-Linnean names like *Royenia* Houst. (for *Loeselia* L.) or *Polemonium* Tournef. (for *Polemonium* L.) have likewise been omitted. Many of the invalid names are, however, listed in the alphabetical index to generic and subgeneric names which follows the section on synonymy. Gray in his paper in 1870 in the Proceedings of the American Academy of Arts and Sciences (vol. 8) used a large number of Latin designations within the genus *Phlox* more or less interchangeably as proper nouns for groups of species and descriptive phrases for key characters. Only those names which we can fairly clearly infer to be section and series names have been included in the present lists.

SYNONYMY

Polemoniaceae

Polemoniaceae (as Polemonia Juss., 1789; Polemonaceae Ventenat, 1799; Polemonideae Pers., 1805; Polemoniaceae DC., 1805; Polemoniariae Reichb., 1837).
Cobaeaceae (as Cobeaceae Don, 1824; Cobaeaceae Dumort., 1829).

A. Tribe Cobaeeae Baill.

Polemoniaceae tribe Cobaeaceae Meisner, Pl. Vasc. Gen. 273, 1839.
Polemoniaceae tribe Cobaeeae Baill., Hist. Pl. 10: 342, 1890[1].
Polemoniaceae tribe Cobaeeae Peter, Nat. Pflzfam. 4 (3a): 44, Dec. 1891[1].
Polemoniaceae subfam. Cobaeoideae Brand, Pflzr. 250: 19, 1907.

1. Cobaea Cav.

Cobaea Cav., Icon. 1 : 11, 1791, not *Cobaea* Necker, 1790; type *Cobaea scandens* Cav.
Rosenbergia Oersted, Vidensk. Meddel. Nat. For. Kjøbenhavn 1856; 30, 1856; type *Rosenbergia gracilis* Oersted (= *Cobaea gracilis* Hemsl.).

(i) Section Cobaea

Cobaea Cav., loc. cit.
Cobaea subg. *Eucobaea* Peter, Nat. Pflzfam. 4 (3a): 44, 1891.
Cobaea sect. *Eucobaea* Pet. ex Brand, Pflzr. 250: 24, 1907.

(ii) Section Aschersoniophila Brand

Cobaea sect. *Aschersoniophila* Brand, Helios 21: 88, 1904; type *Cobaea aschersoniana* Brand.

[1] dates from Flora Malesiana 4 (5), 1954.

(iii) Section Rosenbergia (Oersted) Brand

Rosenbergia Oersted, loc. cit.
Cobaea subg. *Rosenbergia* Peter, Nat. Pflzfam. 4 (3a): 44, 1891.
Cobaea sect. *Rosenbergia* Brand, Pflzr. 250: 28, 1907.

B. Tribe Cantueae Peter

Polemoniaceae tribe Cantueae Peter, Nat. Pflzfam. 4 (3a): 45, 1891.

2. Cantua Juss.

Cantua Juss. ex Lam., Encycl. 1: 603, 1783; type *Cantua buxifolia* Juss.
Periphragmos Ruiz & Pavon, Prodr. Fl. Peruv. 26 & t. 4, 1794; type *Periphragmos flexuosus* Ruiz & Pavon (= *Cantua pyrifolia* Juss.).
Tunaria Ktze., Rev. Gen. Pl. 3 (2): 228, 1898; type *Tunaria albida* Ktze. (= *Cantua pyrifolia* Juss.).

3. Huthia Brand

Huthia Brand, Engl. Bot. Jahrb. 42: 174, 1908; type *Huthia coerulea* Brand.

C. Tribe Bonplandieae Baill.

Polemoniaceae tribe Bonplandieae Baill., Hist. Pl. 10: 342, 1890.

4. Bonplandia Cav.

Bonplandia Cav., Anal. Hist. Nat. 2: 131 & t. 20, 1800, non Willd.; type *Bonplandia geminiflora* Cav.
Caldasia Willd., Hort. Berol. 71 & t. 71, 1807; type *Caldasia heterophylla* Willd. (= *Bonplandia geminiflora* Cav.).

5. Loeselia L.

Loeselia L., Gen. Pl., ed. 5, 276, 1754; type *Loeselia ciliata* L. acc. Int. Rules Bot. Nomencl., ed. 3, Suppl., 1935.
Hoitzia Juss., Gen. Pl. 136, 1789; type *Hoitzia mexicana* Lam. (= *Loeselia mexicana* Brand).

(i) Section Loeselia

Loeselia L., loc. cit.
Hoitzia Juss., loc. cit.
Loeselia subg. *Euloeselia* Peter, Nat. Pflzfam. 4 (3a): 53, 1891.
Loeselia sect. *Euloeselia* Peter ex Brand, Pflzr. 250: 174, 1907.

(ii) *Section Glumiselia V. Grant, sect. nov.*

Frutex ad 2.5 m. altus. Folia opposita, grandia, in inflorescentibus reducta. Inflorescentia paniculata. Bractae purpureae, circum basem floris adpressae. Corolla hypocrateriformis, alba; lobi corollae unguiculati, longi et lineares. Typus Loeselia grandiflora Standley.

Loeselia grandiflora.

The name, compounded from "glume" and "Loeselia", is intended to denote the glume-like appearance of the floral bracts.

D. Tribe Polemonieae Reichb.

Polemoniariae Gruppe Polemonieae Reichb., Handb. Nat. Pflzsystems, ed. 1, 194, 1837.
Polemoniaceae tribe Polemonieae Meisner, Pl. Vasc. Gen. 273, 1839.
Polemoniaceae tribe Polemonieae Baill., Hist. Pl. 10: 339, 1890.
Polemoniaceae tribe Polemonieae Peter, Nat. Pflzfam. 4 (3a): 46, Dec. 1891.
Polemoniaceae subfam. Polemonioideae Brand, Pflzr. 250: 30, 1907.
Polemoniariae Gruppe Phloginae Reichb., loc. cit. 194.

6. Polemonium L.

Polemonium L., Gen. Pl., ed. 5, 76, 1754; type *Polemonium caeruleum* L. acc. Int. Rules Bot. Nomencl., ed. 3, Suppl., 1935.
Polemoniella Heller, Muhlenbergia 1: 57, 1904; type *Polemonium micranthum* Benth.

(i) Section Polemonium

Polemonium L., loc. cit.
Polemonium subg. *Eupolemonium* Peter, Nat. Pflzfam. 4 (3a): 52, 1891.
Polemonium sect. *Eupolemonium* Peter ex Brand, Pflzr. 250: 31, 1907.
Polemonium ser. *Pulchella* Rydb., Fl. Rocky Mts., ed. 1, 680, 1917; type *Polemonium pulcherrimum* Hook.
Polemonium ser. *Caerulea* Rydb., op. cit., 680; type *Polemonium caeruleum* L.
Polemonium sect. *Eupolemonium* ser. *Coerulea* Vassilev, Fl. U.S.S.R. 19: 83, 1953; type *Polemonium caeruleum* L.
Polemonium sect. *Eupolemonium* ser. *Villosa* Vassilev, op. cit., 80; type *Polemonium villosum* Rud. (= *P. caeruleum*).
Polemonium sect. *Eupolemonium* ser. *Humilia* Vassilev, op. cit., 85; type *Polemonium humile* Willd.
Polemonium sect. *Eupolemonium* ser. *Parviflora* Vassilev, op. cit., 92; type *Polemonium parviflorum* Tolm. (near *P. caeruleum*).

(ii) Section *Melliosma* Peter ex Brand

Polemonium subg. *Melliosma* Peter, Nat. Pflzfam. 4 (3a): 52, 1891; type here selected as *Polemonium confertum* Gray.
Polemonium sect. *Melliosma* Peter ex Brand, Pflzr. 250: 43, 1907.
Polemonium ser. *Viscosa* Rydb., Fl. Rocky Mts., ed. 1, 680, 1917; type *Polemonium viscosum* Nutt.

(iii) Section *Polemoniastrum* Peter ex Brand

Polemonium subg. *Polemoniastrum* Peter, Nat. Pflzfam. 4 (3a): 52, 1891; type here selected as *Polemonium micranthum* Benth.
Polemonium sect. *Polemoniastrum* Peter ex Brand, Pflzr. 250: 45, 1907.
Polemonium ser. *Micrantha* Rydb., Fl. Rocky Mts., ed. 1, 680, 1917; type *Polemonium micranthum* Benth.
Polemoniella Heller, loc. cit.

7. Allophyllum (Nutt.) A. & V. Grant

Gilia sect. *Allophyllum* Nutt., Jour. Acad. Nat. Sci. Philadelphia, ser. 2, 1: 155, 1848; type *Gilia divaricata* Nutt. (= *Allophyllum divaricatum* (Nutt.) A. & V. Grant).
Allophyllum A. & V. Grant, El Aliso 3: 98, 1955.
Collomia sect. *Gilioides* Benth., DC. Prodr. 9: 308, 1845; type *Collomia gilioides* Benth. (= *Allophyllum gilioides* (Benth.) A. & V. Grant).
Collomia sect. *Eucollomia* ser. *Giliaeformes* Gray, Proc. Amer. Acad. 8: 260, 1870; type *Collomia gilioides* Benth.
Gilia sect. *Phlogastrum* Brand, Pflzr. 250: 88, 1907; type *Gilia divaricata* Nutt. acc. A. & V. Grant, El Aliso 3: 99, 1955.
Gilia subg. *Greenianthus* Mason & A. Grant, Madroño 9: 206, 1948; type *Collomia gilioides* Benth.

8. Collomia Nutt.

Collomia Nutt., Gen. N. Am. Pl. 1: 126, 1818; type *Collomia linearis* Nutt.
Courtoisia Reichb., Cat. Hort. Dresden 1829 ex Icon. Bot. Exot. 3 : 4, 1829; type *Courtoisia daucifolia* Reichb. (= *Collomia heterophylla* Hook.).

(i) Section *Collomiastrum* Brand

Collomia sect. *Collomiastrum* Brand, Pflzr. 250: 52, 1907; type here selected as *Collomia mazama* Cov.

(ii) Section *Courtoisia* (*Reichb.*) *Wherry*

Courtoisia Reichb., loc. cit.
Gilia sect. *Courtoisia* Gray, Syn. Fl. N. Amer., ed. 2, 2 (1): 408, 1886.
Collomia sect. *Courtoisia* Wherry, Amer. Midl. Nat. 31: 218, 1944.

(iii) Section *Collomia*

Collomia Nutt., loc. cit.
Collomia sect. *Eucollomia* Benth., DC. Prodr. 9: 307, 1845.
Gilia sect. *Collomia* Gray, Syn. Fl. N. Amer., ed. 2, 2 (1) Suppl.: 408, 1886.

9. Gymnosteris Greene

Gymnosteris Greene, Pittonia 3: 303, 1898; type *Collomia nudicaulis* Hook. & Arn. (= *Gymnosteris nudicaulis* (H. & A.) Greene).

10. Phlox L.

Phlox L., Gen. Pl., ed. 5, 75, 1754; type *Phlox glaberrima* L. acc. Int. Rules Bot. Nomencl., ed. 3, Suppl., 1935.
Fonna Adans., Fam. 2: 214, 1763; type here interpreted as *Phlox glaberrima* L.
Armeria L., System. Vegetab., ed. 1, 1735, non Gen. Pl., ed. 1, 1737; type here interpreted as *Phlox glaberrima* L.

(i) Section *Phlox*

Phlox L., loc. cit.
Phlox sect. *Latifoliae* Benth., DC. Prodr. 9: 303, 1845; type here interpreted as *Phlox glaberrima* L.
Phlox sect. *Subulatae* Benth., op. cit. 305; type *Phlox subulata* L.
Phlox sect. *Latifoliae* ser. *Thyrsiflorae* Gray, Proc. Amer. Acad. 8: 249, 1870; type *Phlox paniculata* L.
Phlox sect. *Latifoliae* ser. *Corymosae* Gray, op. cit. 249; type *Phlox glaberrima* L.
Phlox sect. *Latifoliae* ser. *Sparsiflorae* Gray, op. cit. 251; type *Phlox bifida* Beck.
Phlox sect. *Reptantes* Peter, Nat. Pflzfam. 4 (3a): 46, 1891; type *Phlox reptans* Michx. (= *P. stolonifera* Sims).
Phlox sect. *Paniculatae* Peter, op. cit. 46; type *Phlox paniculata* L.
Phlox subg. *Macrophlox* Brand, Pflzr. 250: 58, 1907; type *Phlox glaberrima* L.
Phlox subg. *Macrophlox* sect. *Euphlox* Brand, op. cit. 58.
Phlox subg. *Macrophlox* sect. *Euphlox* subsect. *Holanthium* Brand, op. cit. 58; type here interpreted as *Phlox glaberrima* L.

Phlox subg. *Macrophlox* sect. *Euphlox* subsect. *Meranthium* Brand, op. cit. 72; type here selected as *Phlox bifida* Beck.

Phlox subg. *Macrophlox* sect. *Heterostylon* Brand, op. cit. 77; type *Phlox subulata* L.

Phlox ser. *Longifoliae* Rydb., Fl. Rocky Mts., ed. 1, 684, 1917; type *Phlox longifolia* Nutt.

Phlox sect. *Ovatae* Wherry, Bartonia 13: 18, 1932; type *Phlox ovata* L.

Phlox sect. *α-Phlox* Wherry, The Genus Phlox, 66, 1955; type *Phlox glaberrima* L.

Phlox sect. *α-Phlox* subsect. *Subulatae* Wherry, op. cit. 66.

Phlox sect. *α-Phlox* subsect. *Stoloniferae* Wherry, op. cit. 74; type *Phlox stolonifera* Sims.

Phlox sect. *α-Phlox* subsect. *Cluteanae* Wherry, op. cit. 78; type *Phlox cluteana* A. Nels.

Phlox sect. *α-Phlox* subsect. *Longifoliae* Wherry, op. cit. 81.

Phlox sect. *α-Phlox* subsect. *Ovatae* Wherry, op. cit. 98.

Phlox sect. *α-Phlox* subsect. *Paniculatae* Wherry, op. cit. 117.

(ii) Section Divaricatae Peter

Phlox sect. *Annuae, Texenses* Gray, Proc. Amer. Acad. 8: 257, 1870; type here selected as *Phlox drummondii* Hook.

Phlox sect. *Annuae, Texenses* ser. *Uniovulatae* Gray, op. cit. 257; type *Phlox drummondii* Hook.

Phlox sect. *Annuae, Texenses* ser. *Pluriovulatae* Gray, op. cit. 257; type *Phlox roemeriana* Scheele.

Phlox sect. *Occidentales* ser. *Speciosae* Gray, op. cit. 254; type *Phlox speciosa* Pursh.

Phlox sect. *Divaricatae* Peter, Nat. Pflzfam. 4 (3a): 46, 1891; type *Phlox divaricata* L.

Phlox sect. *Drummondianae* Peter, op. cit. 46; type *Phlox drummondii* Hook.

Phlox subg. *Macrophlox* sect. *Oophila* Brand, Pflzr. 250: 75, 1907; type here selected as *Phlox nana* Nutt.

Phlox sect. *Protophlox* Wherry, The Genus Phlox, 10, 1955; type *Phlox speciosa* Pursh.

Phlox sect. *Protophlox* subsect. *Tenuifoliae* Wherry, op. cit 28; type *Phlox tenuifolia* E. Nels.

Phlox sect. *Protophlox* subsect. *Nanae* Wherry, op. cit. 30; type *Phlox nana* Nutt.

Phlox sect. *Protophlox* subsect. *Divaricatae* Wherry, op. cit. 37.

Phlox sect. *Protophlox* subsect. *Drummondianae* Wherry, op. cit. 58.

(iii) Section Occidentales Gray

Phlox sect. *Occidentales* Gray, Proc. Amer. Acad. 8: 252, 1870; type here selected as *Phlox caespitosa* Nutt.

Phlox sect. *Occidentales* ser. *Pulvinato-caespitosae* Gray, op. cit. 252; type here selected as *Phlox caespitosa* Nutt.

Phlox sect. *Pulvinatae* Peter, Nat. Pflzfam. 4 (3a): 48, 1891; type here selected as *Phlox hoodii* Richardson.

Phlox subg. *Microphlox* Brand, Pflzr. 250: 78, 1907; type *Phlox hoodii* Richardson, acc. Wherry, vide infra.

Phlox subg. *Microphlox* sect. *Neonoma* Brand, op. cit. 79; type here selected as *Phlox kelseyi* Britt.

Phlox subg. *Microphlox* sect. *Chortobolon* Brand, op. cit. 81; type here selected as *Phlox hoodii* Richardson.

Phlox ser. *Multiflorae* Rydb., Fl. Rocky Mts., ed. 1, 684, 1917; type *Phlox multiflora* A. Nels.

Phlox ser. *Caespitosae* Rydb., op. cit. 684; type *Phlox caespitosa* Nutt.

Phlox ser. *Canescentes* Rydb., op. cit. 684; type *Phlox canescens* Torr. & Gray (= *Phlox hoodii* Richardson).

Phlox ser. *Albomarginatae* Rydb., op. cit. 684; type *Phlox albomarginata* Jones.

Phlox sect. *Microphlox* Wherry, The Genus Phlox, 122, 1955.

Phlox sect. *Microphlox* subsect. *Aculeatae* Wherry, op. cit. 122; type *Phlox aculeata* A. Nels.

Phlox sect. *Microphlox* subsect. *Sibiricae* Wherry, op. cit. 125; type *Phlox sibirica* L.

Phlox sect. *Microphlox* subsect. *Albomarginatae* Wherry, op. cit. 132.

Phlox sect. *Microphlox* subsect. *Caespitosae* Wherry, op .cit. 137.

Phlox sect. *Microphlox* subsect. *Douglasianae* Wherry, op. cit. 142; type *Phlox douglasii* Hook.

Phlox sect. *Microphlox* subsect. *Multiflorae* Wherry, op. cit. 146.

Phlox sect. *Microphlox* subsect. *Canescentes* Wherry, op. cit. 152.

11. Microsteris Greene

Microsteris Greene, Pittonia 3: 300, 1898; type *Gilia gracilis* Hook. (= *Microsteris gracilis* (Hook.) Greene).

Gilia subg. *Microsteria* Milliken, Univ. Calif. Publ. Bot. 2: 23, 1904; in part as to type which is *Gilia gracilis* Hook.

Collomia sect. *Myotoca* Griseb., Abh. Königl. Ges. Wissensch. Göttingen 6: 129, 1854, (or p. 41 of a reprint with different pagination); type here selected as *Collomia eritrichioides* Griseb. (= *Microsteris gracilis* (Hook.) Greene).

E. Tribe Gilieae Reichb.

Polemoniariae Gruppe Gilieae Reichb., Handb. Nat. Pflzsystems, ed. 1, 194, 1837.

12. Gilia R. & P.

Gilia R. & P., Prodr. Fl. Peruv. 25 & t. 4, 1794; type *Gilia laciniata* R. & P.

Aliciella Brand, Pflzr. 250: 150, 1907; type *Gilia triodon* Eastw. (= *Gilia leptomeria* Gray).

Giliastrum Rydb., Fl. Rocky Mts., ed. 1, 699 & 1066, 1917; type *Gilia rigidula* Benth.

Tintinabulum Rydb., op. cit. 698 & 1065; type *Gilia filiformis* Parry.

(i) Section Giliastrum Brand

Gilia subg. *Greeneophila* Brand, Pflzr. 250: 144, 1907; type here selected as *Gilia rigidula* Benth.

Gilia subg. *Greeneophila* sect. *Giliastrum* Brand, op. cit. 147; type *Gilia rigidula* Benth.

Giliastrum Rydb., loc. cit.

Gilia subg. *Greeneophila* sect. *Campanulastrum* Brand, Pflzr. 250: 144, 1907; type *Gilia campanulata* Gray.

Gilia subg. *Campanulastrum* Mason & A. Grant, Madroño 9: 219, 1948.

Tintinabulum Rydb., loc. cit.

Gilia subg. *Tintinabulum* Mason & A. Grant, Madroño 9: 220, 1948.

Gilia subg. *Gilmania* Mason & A. Grant, Madroño 9: 205, 1948; type *Gilia latifolia* Wats.

Gilia subg. *Gilia* sect. *Gilmania* V. & A. Grant, El Aliso 3: 299, 1956.

(ii) Section Giliandra Gray

Gilia sect. *Giliandra* Gray, Proc. Amer. Acad. 8: 276, 1870; type here selected as *Gilia stenothyrsa* Gray.

Aliciella Brand, loc. cit.

Gilia ser. *Pinnatifidae* Rydb., Fl. Rocky Mts., ed. 1, 691, 1917; type here selected as *Gilia calcarea* Jones (near *Gilia pinnatifida* Nutt.).

Gilia ser. *Leptomeriae* Rydb., op. cit. 692; type *Gilia leptomeria* Gray.

(iii) Section Gilia

Gilia R. & P., loc. cit.

Gilia sect. *Eugilia* Benth., Bot. Reg. 19: sub t. 1622, 1833.

Gilia subg. *Benthamiophila* Brand, Pflzr. 250: 88, 1907; type here interpreted as *Gilia laciniata* R. & P.

Gilia subg. *Capitata* Milliken, Univ. Calif. Publ. Bot. 2: 23, 1904; type *Gilia capitata* Sims.

Gilia ser. *Capitatae* Rydb., Fl. Rocky Mts., ed. 1, 691, 1917; type *Gilia capitata* Sims.

(*iv*) Section Arachnion A. & V. Grant

Gilia sect. *Arachnion* A. & V. Grant, El Aliso 3: 214, 1956.

Gilia ser. *Inconspicuae* Rydb., Fl. Rocky Mts., ed. 1, 691, 1917; type *Ipomopsis inconspicua* Smith (= *Gilia inconspicua* (Smith) Sweet).

(*v*) Section Saltugilia V. & A. Grant

Gilia sect. *Saltugilia* V. & A. Grant, El Aliso 3: 84, 1954; type *Gilia splendens* Dougl.

Gilia subg. *Kelloggia* Mason & A. Grant, Madroño 9: 219, 1948; type *Gilia capillaris* Kell.

13. Ipomopsis Michx.

Ipomopsis Michx., Fl. Bor. Am., ed. 1, 1: 141, 1803; type *Ipomopsis elegans* Michx. (= *Ipomopsis rubra* (L.) Wherry).

Ipomeria Nutt., Gen. N. Am. Pl. 1: 124, 1818; type *Cantua coronopifolia* Willd. (= *Ipomopsis rubra* (L.) Wherry).

Batanthes Raf., Atl. Jour. 1: 145, 1832; type *Cantua aggregata* Pursh (= *Ipomopsis aggregata* (Pursh) Grant), acc. Grant, El Aliso 3: 353, 1956.

Callisteris Greene, Leafl. Bot. 1: 159, 1905; type *Cantua aggregata* Pursh.

(*i*) Section Phloganthea (Gray) V. Grant

Collomia sect. *Phloganthea* Gray, Proc. Amer. Acad. 8: 260, 1870; type *Phlox pinnata* Cav. (= *Ipomopsis pinnata* (Cav.) Grant), acc. Grant, El. Aliso 3: 353, 1956.

Ipomopsis sect. *Phloganthea* V. Grant, El Aliso 3: 353, 1956.

Loeselia sect. *Giliopsis* Gray, Proc. Amer. Acad. 11: 86, 1876; type *Loeselia tenuifolia* Gray (= *Ipomopsis tenuifolia* (Gray) Grant), acc. Grant, El Aliso 3: 353, 1956.

Loeselia subg. *Giliopsis* Gray ex Peter, Nat. Pflzfam. 4 (3a): 54, 1891.

(*ii*) Section Ipomopsis

Ipomopsis Michx, loc. cit.

Ipomeria Nutt., loc. cit.

Batanthes Raf., loc. cit.

Callisteris Greene, loc. cit.

Gilia sect. *Ipomopsis* Benth., Bot. Reg. 19: sub t. 1622, 1833.

Gilia subg. *Ipomopsis* Milliken, Univ. Calif. Publ. Bot. 2: 24, 1904.

Gilia ser. *Aggregatae* Rydb., Fl. Rocky Mts., ed. 1, 691, 1917; type *Cantua aggregata* Pursh (= *Ipomopsis aggregata* (Pursh) Grant).

Gilia ser. *Longiflorae* Rydb., op. cit. 691; type *Cantua longiflora* Torr. (= *Ipomopsis longiflora* (Torr.) Grant).

(iii) Section Microgilia (Benth.) V. Grant

Gilia sect. *Microgilia* Benth., DC. Prodr. 9: 315, 1845; type *Gilia minutiflora* Benth. (= *Ipomopsis minutiflora* (Benth.) Grant).

Ipomopsis sect. *Microgilia* V. Grant, El Aliso 3: 357, 1956.

Collomia sect. *Picracolla* Nutt., Jour. Acad. Nat. Sci. Philadelphia, ser. 2, 1: 159, 1848; type *Collomia linoides* Nutt. (= *Ipomopsis minutiflora* (Benth.) Grant).

Gilia sect. *Elaphocera* Nutt., op. cit. 155; type *Gilia congesta* Hook. (= *Ipomopsis congesta* (Hook.) Grant), acc. Grant, El Aliso 3: 357, 1956.

Gilia subg. *Elaphocera* Milliken, Univ. Calif. Publ. Bot. 2: 24, 1904.

Navarretia subg. *Hugelia* sect. *Langloisiastrum* Brand, Pflzr, 250: 168, 1907; type *Gilia wrightii* Gray (near *Ipomopsis frutescens* (Rydb.) Grant).

Gilia ser. *Congestae* Rydb., Fl. Rocky Mts., ed. 1, 690, 1917; type *Gilia congesta* Hook. (= *Ipomopsis congesta* (Hook.) Grant).

Gilia sect. *Elaphocera* subsect. *Congestae* Const. & Rollins, Amer. Jour. Bot. 23: 434, 1936; type *Gilia congesta* Hook. (= *Ipomopsis congesta* (Hook.) Grant).

Gilia ser. *Pumilae* Rydb., Fl. Rocky Mts., ed. 1, 691, 1917; type *Gilia pumila* Nutt. (= *Ipomopsis pumila* (Nutt.) Grant).

Gilia ser. *Minutiflorae* Rydb., op. cit. 692; type *Gilia minutiflora* Benth. (= *Ipomopsis minutiflora* (Benth.) Grant).

14. Eriastrum Woot. & Standl.

Hugelia Benth.; Bot. Reg. 19: sub. t. 1622, 1833, not *Hugelia* Reichb., 1828; type here selected as *Hugelia densifolia* Benth. (= *Eriastrum densifolium* (Benth.) Mason).

Welwitschia Reichb., Handb. 194, 1837, not *Welwitschia* Hook., 1862, nom. cons.; type here interpreted as *Hugelia densifolia* Benth.

Eriastrum Woot. & Standl., Contrib. U.S. Natl. Herb. 16: 160, 1913; type *Gilia filifolia* Nutt. (= *Eriastrum filifolium* (Nutt.) Woot. & Standl. ,

Gilia sect. *Collomioides* Endl., Gen. Pl. 657, 1839; type here interpreted as *Hugelia densifolia* Benth.

Gilia sect. *Pseudocollomia* Benth., DC. Prodr. 9: 311, 1845; type *Gilia lutescens* Steud. (= *Eriastrum luteum* (Benth.) Mason).

Gilia sect. *Hugelia* Gray, Proc. Amer. Acad. 8: 271, 1870.

Gilia subg. *Hugelia* Milliken, U.C. Publ. Bot. 2: 24, 1904.

Navarretia subg. *Hugelia* Brand; Pflzr. 250: 164, 1907.

Navarretia subg. *Hugelia* sect. *Euhugelia* Brand, op. cit. 164.

15. Langloisia Greene

Langloisia Greene, Pittonia 3: 30, 1896; type *Navarretia setosissima* Torr. & Gray. Greene listed three species, *L. matthewsii*, *L. schottii* and *L. setosissima*. According to Mason (in Abrams Fl., 1951) the type is *L. schottii*. In 1907 however Brand circumscribed the typical section *Eulangloisia* so as to exclude *L. schottii* but include *L. lanata*, *L. punctata* and *L. setosissima*. Since according to the Rules of Nomenclature (ed. 4, 1952), a lectotype must be selected from among the entities considered by an author, and there is only one element in Brand's *Eulangloisia* which was also considered by Greene, namely *L. setosissima*, we may interpret Brand's action as tantamount to designating that species as the type. The Rules of Nomenclature further prescribe that the first choice of a lectotype should be followed by subsequent workers, which in this case means that Brand's designation takes precedence over that of Mason.

Gilia sect. *Chaetogilia* Gray, Syn. Fl. N. Amer., ed. 2,2 (1, Suppl.): 409, 1886; type here selected as *Loeselia matthewsii* Gray (= *Langloisia matthewsii* (Gray) Greene).

Loeselia subg. *Chaetogilia* Gray ex Peter, Nat. Pflzfam. 4 (3a): 54, 1891.

Gilia subg. *Langloisia* Milliken, Univ. Calif., Publ. Bot. 2: 25, 1904.

Langloisia sect. *Loeseliastrum* Brand, Pflzr. 250: 171, 1907; type here selected as *Loeselia matthewsii* Gray (= *Langloisia matthewsii* (Gray) Greene).

Langloisia sect. *Eulangloisia* Brand, op. cit., 169.

16. Navarretia R. & P.

Navarretia R. & P., Prodr. Fl. Peruv. 20, 1794; type *Navarretia involucrata* R. & P.

Navarretia Hedw., Gen. Pl. 107, 1806; type here interpreted as *Navarretia involucrata* R. & P.

Aegochloa Benth., Bot. Reg. 19: sub t. 1622, 1833; type here selected as *Aegochloa atractyloides* Benth. (= *Navarretia atractyloides* (Benth.) H. & A.).

(i) Section Aegochloa (Benth.) V. Grant, comb. nov

Aegochloa Benth., loc. cit.

Navarretia subg. *Echinocephala* Brand, Pflzr. 250: 152, 1907; type here selected as *Hoitzia squarrosa* Eschsch. (= *Navarretia squarrosa* (Eschsch.) H. & A.).

Navarretia subg. *Echinocephala* sect. *Eunavarretia* Brand, op. cit. 152; does not include *Navarretia involucrata* and so cannot be used as the typical section; the type is here selected as *Navarretia squarrosa*.

(ii) Section Masonia V. Grant, sect. nov.

Plantae mediae ad altae de magnitudinis, virga centrali brevi, base ramosis; foliis ramisque juvenibus glandulosis ad subtiliter glandulosis-puberulis; foliis caulis mediis pinnatis, linearibus, lobis paucis laxisque, spinis non rigidis; bractis cum base lato crasso spinisque rigidis; sinibus calycis non cristatis; corolla regulari, colorata, purpura lutea vel rosea; lobis corollae 3-venosis; staminibus in faucibus corollae insertis; stigmate 3-fissa; capsula dehiscens, basis circumscissilo vel non, 3-loculari, pericarpio coriaceo vel chartaceo; seminibus multis ad paucis in capsula. Typus Gilia breweri Gray.

Navarretia breweri (Gray) Greene; *N. divaricata* (Torr.) Greene; *N. peninsularis* Greene; *N. prolifera* Greene; *N. viscidula* Benth.

Named for Herbert L. Mason, long-time student of the genus.

(iii) Section Mitracarpium Brand

Navarretia subg. *Echinocephala* sect. *Mitracarpium* Brand, Pflzr. 250: 160, 1907; type *Navarretia mitracarpa* Greene.

(iv) Section Navarretia

Navarretia R. & P., loc. cit., not sect. *Eunavarretia* Brand.

Gilia sect. *Navarretia* Endl., Gen. Pl. 657, 1840.

Gilia sect. *Navarretia* Gray, Proc. Amer. Acad. 8: 268, 1870.

Navarretia subg. *Echinocephala* sect. *Baccocarpium* Brand, Pflzr. 250: 161, 1907; type here selected as *Gilia prostrata* Gray (= *Navarretia prostrata* (Gray) Greene).

Navarretia sect. *Fragiles* Crampton, Madroño 12: 235, 1954; type here selected as *Navarretia leucocephala* Benth.

17. Leptodactylon H. & A.

Leptodactylon H. & A., Bot. Beechey's Voy. 369, 1839; type *Leptodactylon californicum* H. & A.

Gilia sect. *Leptodactylon* Benth., DC. Prodr. 9: 316, 1845.

Gilia subg. *Leptodactylon* Milliken, Univ. Calif. Publ. Bot 2: 24, 1904.

18. Linanthus Benth.

Linanthus Benth., Bot. Reg. 19: sub t. 1622, 1833; type *Linanthus dichotomus* Benth.

Leptosiphon Benth., *loc. cit.;* type here selected as *Leptosiphon androsaceus* Benth. (= *Linanthus androsaceus* (Benth.) Greene).

Fenzlia Benth., *loc. cit.;* type Fenzlia dianthiflora Benth. (= *Linanthus dianthiflorus* (Benth.) Greene).

Dactylophyllum Spach, Hist. Vég. Phan. 9: 108, 1840, not *Dactylophyllum* Spenn, 1829; type *Linanthus liniflorus* (Benth.) Greene, acc. Grant, vide infra.

Siphonella Nutt. ex Gray, Proc. Amer. Acad. 8: 267, 1870, as a synonym.

Siphonella Heller, Muhlenbergia 8: 57, 1912, not *Siphonella* Small, 1903; type *Gilia nuttallii* Gray. (= *Linanthus nuttallii* (Gray) Greene), acc. Grant, vide infra.

Linanthastrum Ewan, Jour. Wash. Acad. Sci. 32: 139, 1942; type *Gilia nuttallii* Gray (= *Linanthus nuttallii* (Gray) Greene).

(i) Section Siphonella (Gray) V. Grant, comb. nov.

Gilia sect. *Siphonella* Gray, Proc. Amer. Acad. 8: 266, 1870; type here selected as *Gilia nuttallii* Gray (= *Linanthus nuttallii* (Gray) Greene).

Siphonella Nutt. ex Gray, loc. cit.

Siphonella Heller, loc. cit.

Linanthastrum Ewan, loc. cit.

(ii) Section Pacificus (Jeps.) V. Grant, comb. nov.

Linanthus subg. *Pacificus* Jeps., Man. Fl. Pl. Calif. 800, 1925; type here interpreted as *Linanthus pacificus* Milliken.

(iii) Section Leptosiphon (Endl.) V. Grant, comb. nov.

Leptosiphon Benth., loc. cit.

Gilia sect. *Leptosiphon* Endl., Gen. Pl. 657, 1840.

Linanthus subg. *Leptosiphon* Jeps., Fl. W. Mid. Calif., 429, 1901.

(*iv*) *Section Dactylophyllum (Benth.) V. Grant, comb. nov.*

Gilia sect. *Dactylophyllum* Benth., Bot. Reg. 19: sub t. 1622, 1833; type here
   selected as *Gilia liniflora* Benth. (= *Linanthus liniflorus* (Benth.) Greene).
Dactylophyllum Spach, loc. cit.
Gilia sect. *Dactylophyllum* Gray, Proc. Amer. Acad. 8: 263, 1870.
Linanthus subg. *Dactylophyllum* Jeps., Fl. W. Mid. Calif., 429, 1901.
Gilia sect. *Chrysantha* Nutt., Jour. Acad. Nat. Sci. Philadelphia, ser. 2, 1: 155,
   1848; type *Gilia aurea* Nutt. (= *Linanthus aureus* (Nutt.) Greene).

(*v*) *Section Linanthus*

Linanthus Benth., loc. cit.
Gilia sect. *Eulinanthus* Endl., Gen. Pl. 657, 1840.
Linanthus subg. *Eulinanthus* Jeps., Fl. W. Mid. Calif., 429, 1901.
Gilia subg. *Grayophila* Brand, Pflzr. 250: 124, 1907; type here interpreted
   as *Linanthus dichotomus* Benth.
Linanthus subg. *Millikenia* Jeps., Man. Fl. Pl. Calif. 800, 1925; type *Linanthus*
   concinnus Milliken.

(*vi*) *Section Dianthoides (Endl.) V. Grant, comb. nov.*

Gilia sect. *Dianthoides* Endl., Gen. Pl. 657; type here interpreted as *Fenzlia*
   dianthiflora Benth. (= *Linanthus dianthiflorus* (Benth.) Greene).
Fenzlia Benth., loc. cit.
Linanthus subg. *Fenzlia* Jeps., Man. Fl. Pl. Calif. 800, 1925.
Gilia sect. *Fornicogilia* Peter, Nat. Pflzfam. 4 (3a): 51, 1891; type *Gilia*
   parryae Gray (= *Linanthus parryae* (Gray) Greene).
Linanthus subg. *Parrya* Jeps., Man. Fl. Pl. Calif. 800, 1925; type *Gilia parryae*
   Gray.

Aculeatae. Phlox iii.
Aegochloa. Navarretia.
Aggregatae. Ipomopsis ii.
Albomarginatae. Phlox iii.
Aliciella. Gilia.
Allophyllum. genus 7.
Alpha-Phlox. Phlox i.
Annuae, Texenses. Phlox ii.
Arachnion. Gilia iv.
Armeria. Phlox.
Aschersoniophila. Cobaea ii.
Baccocarpium. Navarretia iv.
Batanthes. Ipomopsis.
Benthamiophila. Gilia iii.
Bonplandia. genus 4.
Brickellia. nomen nudum of Rafinesque,
  cited by older authors in synonymy
  under Ipomopsis.
Caerulea. Polemonium i.
Caespitosae. Phlox iii.
Caldasia. Bonplandia.
Callisteris. Ipomopsis.
Callomia. Collomia.
Campanulastrum. Gilia i.
Canescentes. Phlox iii.
Cantua. genus 2.
Capitata. Gilia iii.
Capitatae. Gilia iii.
Chaetogilia. Langloisia.
Chortobolon. Phlox iii.
Chrysantha. Linanthus iv.
Cluteanae. Phlox i.
Cobaea. genus 1.
Cobbea. Cobaea
Cobea. Cobaea.
Coerulea. Polemonium i.
Collomia. genus 8.
Collomiastrum. Collomia i.
Collomioides. Eriastrum.
Congestae. Ipomopsis iii.
Corymbosae. Phlox i.
Courtoisia. Collomia.
Cullomia. Collomia.
Curtoisia. Courtoisia.
Dactylophyllum. Linanthus.
Dianthoides. Linanthus vi.
Divaricatae. Phlox ii.
Douglasianae. Phlox iii.
Drummondianae. Phlox ii.

Dupratzia. old name of Rafinesque
  formerly attributed to Phlox but now
  identified as a synonym of Eustoma,
  Gentianaceae (cf. Wherry, Castanea 20:
  71, 1955).
Echinocephala. Navarretia i.
Elaphocera. Ipomopsis iii.
Eriastrum. genus 14.
Eucobaea. Cobaea i.
Eucollomia. Collomia iii.
Eugilia. Gilia iii.
Euhugelia. Eriastrum.
Eulangloisia. Langloisia.
Eulinanthus. Linanthus v.
Euloeselia. Loeselia i.
Eunavarretia. Navarretia i.
Euphlox. Phlox i.
Eupolemonium. Polemonium i.
Fentzlia. Fenzlia.
Fenzlia. Linanthus.
Fonna. Phlox.
Fornicogilia. Linanthus vi.
Fragiles. Navarretia iv.
Gilia. genus 12.
Giliaeformes. Allophyllum.
Giliandra. Gilia ii.
Giliastrum. Gilia.
Gilioides. Allophyllum.
Giliopsis. Ipomopsis i.
Gilmania. Gilia i.
Glumiselia. Loeselia ii.
Grayophila. Linanthus v.
Greenianthus. Allophyllum.
Greeneophila. Gilia i.
Gymnosteris. genus 9.
Heterostylon. Phlox i.
Hoitzia. Loeselia i.
Holanthium. Phlox i.
Huegelia. Hugelia.
Hugelia. Eriastrum.
Humilia. Polemonium i.
Huthia. genus 3.
Inconspicuae. Gilia iv.
Ipomeria. Ipomopsis.
Ipomopsis. genus 13.
Kelloggia. Gilia v.
Langloisia. genus 15.
Langloisiastrum. Ipomopsis iii.
Latifoliae. Phlox i.

Leptodactylon. genus 17.
Leptomeriae. Gilia ii.
Leptosiphon. Linanthus.
Linanthastrum. Linanthus.
Linanthus. genus 18.
Loeselia. genus 5.
Loeseliastrum. Langloisia.
Longiflorae. Ipomopsis ii; also misspel-
     ling of Longifoliae.
Longifoliae. Phlox i.
Lychnidea. pre-Linnean name of Plukenet
     and Dillenius for Phlox.
Lychnoides. pre-Linnean name of Ray
     for Phlox.
Macrophlox. Phlox i.
Masonia. Navarretia ii.
Melliosma. Polemonium ii.
Meranthium. Phlox i.
Micrantha. Polemonium iii.
Microgilia. Ipomopsis iii.
Microphlox. Phlox iii.
Microsteria. Microsteris.
Microsteris. genus 11.
Millikenia. Linanthus v.
Minutiflorae. Ipomopsis iii.
Mitracarpium. Navarretia iii.
Multiflorae. Phlox iii.
Myotoca. Microsteris. as a generic name
     Myotoca Griseb. is a nomen nudum.
Nanae. Phlox ii.
Navarretia. genus 16.
Neonoma. Phlox iii.
Occidentales. Phlox iii.
Ogochloa. Aegochloa.
Oophila. Phlox ii.
Ovatae. Phlox i.
Pacificus. Linanthus ii.
Paniculatae. Phlox i.
Parrya. Linanthus vi.
Parviflora. Polemonium i.
Periphragmos. Cantua.

Phloganthea. Ipomopsis i.
Phlogastrum. Allophyllum.
Phlox. genus 10.
Phloxus. Phlox.
Phu. pre-Linnean name of Dodonaeus,
     Gesner and Ruppius for Polemonium.
Picracolla. Ipomopsis iii.
Pinnatifidae. Gilia ii.
Pluriovulatae. Phlox ii.
Polemoniastrum. Polemonium iii.
Polemoniella. Polemonium.
Polemonium. genus 6.
Protophlox. Phlox ii.
Pseudocollomia. Eriastrum.
Pulchella. Polemonium i.
Pulvinatae. Phlox iii.
Pulvinato-caespitosae. Phlox iii.
Pumilae. Ipomopsis iii.
Reptantes. Phlox i.
Rosenbergia. Cobaea.
Rossmaesslera. nomen nudum of
     Reichenbach, cited by older authors in
     synonymy under Gilia.
Royenia. pre-Linnean name of Houstoun
     for Loeselia.
Saltugilia. Gilia v.
Sibiricae. Phlox iii.
Siphonella. Linanthus.
Sparsiflorae. Phlox i.
Speciosae. Phlox ii.
Stoloniferae. Phlox i.
Subulatae. Phlox i.
Tenuifoliae. Phlox ii.
Thyrsiflorae. Phlox i.
Tintinabulum. Gilia.
Tunaria. Cantua.
Uniovulatae. Phlox ii.
Villosa. Polemonium i.
Viscosa. Polemonium ii.
Welwitschia. Eriastrum.

Species accepted in the present work are listed in italics and indexed as to genus number and section number. The other binomials in roman type are given as synonyms of the accepted species. Synonymy for the purpose of this list comprises three distinct situations: out and out synonymy, tentative reference of one entity to another pending further study, and inclusion of an entity in a polytypic species in the rank of subspecies or variety.

Some names have had an interesting history of travels. Thus *rubra* L. has been placed successively in the genera *Polemonium, Ipomoea, Navarretia, Gilia,* and *Ipomopsis; gracilis* Dougl. has been in *Collomia, Gilia, Eutoca, Navarretia, Phlox,* and *Microsteris; pinnata* Cav. in *Phlox, Gilia, Navarretia,* and *Ipomopsis; pungens* Torr. in *Cantua, Batanthes, Gilia, Navarretia,* and *Leptodactylon; gilioides* Benth. in *Collomia, Gilia, Navarretia, Microsteris,* and *Allophyllum;* and *nuttallii* Gray in *Gilia, Navarretia, Linanthus, Leptodactylon, Siphonella,* and *Linanthastrum.*

It has often happened that the same binomial was created by different authors for entirely different plants. Our list then shows *Planta specifica* Smith as equivalent to one entity and *Planta specifica* Brown as equivalent to another. A different situation arises when *P. specifica* Smith has been used in different senses by subsequent authors. The various taxonomic indices have given formal recognition to many such usages·by including "*P. specifica* Brown" and "*P. specifica* Adams" in addition to the original *P. specifica* Smith. Names of this sort which are not homonyms but merely different interpretations of a single binomial are omitted from the present list, even though they have been included in other compendia.

It will be obvious to anyone who is acquainted with the taxonomic status of many groups within the family that the designation of acceptable species and of their unaccepted equivalents must require numerous fairly arbitrary decisions. As it was not my purpose in this work to produce a critical revision of the family at the species level, the assignments are in many cases only approximate. The following list is an aid to, not a substitute for, future revisional studies.

Aegochloa

atractylioides Bentham	Navarretia atractyloides
cotulaefolia Bentham	Navarretia cotulaefolia
eryngioides Bentham	Navarretia involucrata
intertexta Bentham	Navarretia intertexta
pubescens Bentham	Navarretia pubescens
pungens Bentham	Navarretia squarrosa.
torreyi Don	Leptodactylon pungens

Aliciella

triodon (Eastwood) Brand	Gilia leptomeria

Allophyllum

divaricatum (Nuttall) Grant & Grant	7
gilioides (Bentham) Grant & Grant	7
glutinosum (Bentham) Grant & Grant	7
integrifolium (Brand) Grant & Grant	7
violaceum (Heller) Grant & Grant	7

Armeria

miscellaneous species	Armeria L. (Plumbaginaceae)
adsurgens (Torrey ex Gray) Kuntze	Phlox adsurgens
amoena (Sims) Kuntze	Phlox amoena
bifida (Beck) Kuntze	Phlox bifida
bryodes Kuntze	see A. bryoides
bryoides (Nuttall) Kuntze	Phlox bryoides
caespitosa (Nuttall) Kuntze	Phlox caespitosa
canescens (Torrey & Gray) Kuntze	Phlox hoodii.
divaricata (L.) Kuntze	Phlox divaricata
douglasii (Hooker) Kuntze	Phlox douglasii
drummondii (Hooker) Kuntze	Phlox drummondii.
floridana (Bentham) Kuntze	Phlox floridana
glaberrima (L.) Kuntze	Phlox glaberrima
hoodii (Richardson) Kuntze	Phlox hoodii
hordii Kuntze	see A. hoodii
linearifolia (Gray) Kuntze	Phlox longifolia
longifolia (Nuttall) Kuntze	Phlox longifolia
maculata (L.) Kuntze	Phlox maculata
muscodes Kuntze	see A. muscoides
muscoides (Nuttall) Kuntze	Phlox hoodii
nana (Nuttall) Kuntze	Phlox nana
ovata (L.) Kuntze	Phlox ovata
paniculata (L.) Kuntze	Phlox paniculata
pilosa (L.) Kuntze	Phlox pilosa
reptans (Michaux) Kuntze	Phlox stolonifera
richardsoni (Hooker) Kuntze	Phlox richardsoni
roemeriana (Scheele) Kuntze	Phlox roemeriana
sibirica (L.) Kuntze	Phlox sibirica
speciosa (Pursh) Kuntze	Phlox speciosa
stellaria (Gray) Kuntze	Phlox bifida
subulata (L.) Kuntze	Phlox subulata

Batanthes

aggregata Rafinesque	Ipomopsis aggregata
agregata Rafinesque	see B. aggregata

...(*continued*)

Batanthes

arizonica Greene	Ipomopsis aggregata
attenuata Greene	Ipomopsis aggregata
bridgesii Greene	Ipomopsis aggregata
collina Greene	Ipomopsis aggregata
flavida Greene	Ipomopsis aggregata
formosissima Greene	Ipomopsis aggregata
leucantha Greene	Ipomopsis aggregata
longiflora Rafinesque	Ipomopsis longiflora
pulchella Greene	Ipomopsis aggregata
pungens Rafinesque	Leptodactylon pungens
scopulorum Greene	Ipomopsis aggregata
texana Greene	Ipomopsis aggregata

Bonplandia

geminiflora Cavanilles	4
linearis Robinson	4

Caldasia

miscellaneous species	Oremyrrhis (Umbelliferae) or Helosis (Balanophoraceae)
heterophylla Willdenow	Bonplandia geminiflora

Callisteris

aggregata Greene	Ipomopsis aggregata
arizonica Greene	Ipomopsis aggregata
attenuata Greene	Ipomopsis aggregata
bridgesii Greene	Ipomopsis aggregata
collina Greene	Ipomopsis aggregata
flavida Greene	Ipomopsis aggregata
formosissima Greene	Ipomopsis aggregata
leucantha Greene	Ipomopsis aggregata
pulchella Greene	Ipomopsis aggregata
texana Greene	Ipomopsis aggregata
violacea Greene	Ipomopsis tenuituba

Cantua

aggregata Pursh	Ipomopsis aggregata
bicolor Lemaire	2
breviflora Jussieu	Gilia laciniata
brevifolia Steudel	see C. breviflora
buxifolia Jussieu ex Lamarck	2
candelilla Brand	2
coccinea Poiret ex Lamarck	Gilia laciniata
coerulea Poiret ex Lamarck	Loeselia mexicana
cordata Jussieu,	nomen dubium
coronopifolia Willdenow	Ipomopsis rubra
cuneifolia Jussieu ex Roemer & Schultes	nomen dubium
dependens Persoon	C. buxifolia
elegans Poiret	Ipomopsis rubra
fasciculata Willdenow ex Roemer & Schultes	Fouquieria spinosa (Fouquieriaceae)
flexuosa Persoon	C. pyrifolia
floridana Nuttall	Ipomopsis rubra
foetida Persoon	Vestia lycioides (Solanaceae)

...(*continued*)

Cantua

glandulosa Poiret	Loeselia glandulosa
glomeriflora Jussieu	Ipomopsis pinnata
glutinosa Presl	nomen dubium, Gilia crassifolia?
hibrida Herrera	see C. hybrida
hoitzia Willdenow	Loeselia mexicana
hybrida Herrera,	nomen dubium, probably not Cantua, perhaps not Polemoniaceae
laciniata Poiret ex Lamarck	Gilia laciniata
lanceolata Peter	C. buxifolia
ligustrifolia Jussieu	Vestia lycioides (Solanaceae)
longiflora Torrey	Ipomopsis longiflora
longifolia Brand	near C. pyrifolia
loxensis Willdenow	C. pyrifolia
megapotamica Sprengel	nomen dubium, probably not Polemoniaceae
ochroleuca Brand	2
ovata Cavanilles	C. buxifolia
parviflora Pursh	Gilia inconspicua
peruviana Gmelin	C. pyrifolia
picta Hort	Ipomopsis rubra
pinnatifida Lamarck	Ipomopsis rubra
pirifolia Jussieu	see C. pyrifolia
pungens Torrey	Leptodactylon pungens
pyrifolia Jussieu	2
quercifolia Jussieu	2
sinuata Willdenow ms	C. quercifolia
theafolia Don	C. buxifolia
thyrsoidea Jussieu	Ipomopsis rubra
tomentosa Cavanilles	C. buxifolia
tuberosa Roemer & Schultes	Ipomoea muricata (Convolvulaceae)
uniflora Persoon	C. buxifolia

Cobaea

acuminata De Candolle	C. lutea
aequatoriensis Asplund	1–iii
aschersoniana Brand	1–ii
biaurita Standley	1–i
campanulata Hemsley	1–i
equatoriensis, Gray Index	see C. aequatoriensis
gracilis (Oersted) Hemsley	1–iii
hookeriana Standley	1–iii
lasseri Pittier	1–i
lutea Don	1–i
macrostema Pavon	C. lutea
macrostoma De Candolle	see C. macrostema
minor Martens & Galeotti	1–i
pachysepala Standley	1–i
panamensis Standley	1–i
penduliflora Hooker	C. hookeriana
penduliflora (Karsten) Hooker	1–iii
pringlei (House) Standley	1–i
scandens Cavanilles	1–i
skutchii Johnston	1–i
steyermarkii Standley	near C. viorna
stipularis Bentham	1–i
tomentulosa Standley	near C. pachysepala

...*(continued)*

Cobaea

trianaei Hemsley	1–i
triflora Smith	1–i
villosa Standley	near C. viorna
viorna Standley	1–i

Collomia

aggregata (Pursh) Porter	Ipomopsis aggregata
aristella (Gray) Rydberg	C. tinctoria
atacamensis Philippi	nomen dubium, perhaps Gilia or perhaps not Polemoniaceae
bellidifolia Douglas ex Hooker	Microsteris gracilis
biflora (Ruiz & Pavon) Brand	8–iii
cavanillesiana Don	Ipomopsis pinnata
cavanillesii Hooker & Arnott.	8–iii
chubutensis Spegazzini	Microsteris gracilis?
coccinea Lehmann ex Lindley	C. cavanillesii
debilis (Watson) Greene	8–i
diversifolia Greene	8–ii
dulcis Lindley	Collania (Bomarea) dulcis (Amaryllidaceae)
eritrichioides Grisebach	Microsteris gracilis
erythraeoides Grisebach	Microsteris gracilis
gilioides Bentham	Allophyllum gilioides
giliopsis Smyth	see C. gilioides
glutinosa Bentham	Allophyllum glutinosum
gracilis Douglas ex Hooker	Microsteris gracilis
grandiflora Douglas ex Lindley	8–iii
heterophylla Hooker	8–ii
howardi Jones	C. debilis
humilis Douglas ex Hooker	Microsteris gracilis
hurdlei Nelson	C. debilis
lanceolata Greene ex Brand	C. linearis Nuttall
larseni (Gray) Payson	8–i
lateritia Don in Sweet	C. cavanillesii
leptalea Gray	Gilia leptalea
linearis (Cavanilles) Nuttall.	C. cavanillesii
linearis Nuttall	8–iii
linoides Nuttall	Ipomopsis minutiflora
longiflora (Torrey) Gray	Ipomopsis longiflora
macrocalyx Leiberg ex Brand	8–iii
mazama Coville	8–i
micrantha Kellogg.	Microsteris gracilis
myotoca, Index Kewensis	Microsteris gracilis
navarretia Don	Navarretia involucrata
nudicaulis Hooker & Arnott	Gymnosteris nudicaulis
parviflora Hooker	C. linearis Nuttall
patagonica Spegazzini	Nicotiana linearis (Solanaceae)
pringlei (Gray) Peter	near Ipomopsis macombii.
pusilla Dusén	Androsace salasi (Primulaceae)
rawsoniana Greene	8–i
scabra Greene	C. grandiflora
setosa Sieber ex Steudel	Felicia reflexa (Compositae)
sinistra (Jones) Brand	Gilia capillaris
soehrensi Philippi	C. cavanillesii
stenosiphon Kuntze	C. cavanillesii

...(*continued*)

Collomia

tenella Gray	8–iii
thurberi Gray	Ipomopsis thurberi
tinctoria Kellogg	8–iii
tracyi Mason	8–iii
unidentata Bertero ex Brand	C. cavanillesii

Courtoisia

miscellaneous species	Courtoisia Nees (Cyperaceae)
bipinnatifida Reichenbach	Collomia heterophylla
daucifolia Reichenbach	Collomia heterophylla

Dactylophyllum

ambiguum Heller	Linanthus ambiguus

Eriastrum

abramsii (Elmer) Mason	14
brandegeae Mason	14
densifolium (Bentham) Mason	14
diffusum (Gray) Mason	14
eremicum (Jepson) Mason	14
filifolium (Nuttall) Wooton & Standley	14
hooveri (Jepson) Mason	14
luteum (Bentham) Mason	14
pluriflorum (Heller) Mason	14
sapphirinum (Eastwood) Mason	14
sparsiflorum (Eastwood) Mason	14
tracyi Mason	14
virgatum (Bentham) Mason	14
wilcoxii (Nelson) Mason	14

Eutoca

miscellaneous species	Eutoca Brown (Phacelia, Hydrophyllaceae)
gracilis Grisebach ms.	Microsteris gracilis

Fenzlia

miscellaneous species	Fenzlia Endl. (Myrtaceae)
concinna Nuttall	Linanthus dianthiflorus
dianthiflora Bentham	Linanthus dianthiflorus
speciosa Nuttall	Linanthus dianthiflorus

Fonna

andicola Lunell	Phlox andicola
hoodii Nieuwland & Lunell	Phlox hoodii
kelseyi Nieuwland & Lunell	Phlox kelseyi

Gilia

abramsii Mason & Grant	G. ochroleuca
abrotanifolia Nuttall ex Greene	G. capitata
achilleaefolia Bentham	12–iii
acerosa (Gray) Britton	G. rigidula
aggregata (Pursh) Sprengel	Ipomopsis aggregata
aliquanta Grant & Grant	12–iv
alba Hort.	G. tricolor
alpina (Weddell) Brand	G. laciniata

...(*continued*)

Gilia

alpina Eastwood	G. leptantha
ambigua Rattan	Linanthus ambiguus
andicola Philippi	G. crassifolia
androsacea (Bentham) Steudel	Linanthus androsaceus
angelensis Grant	12–iii
arcuata Hieronymus	Polemonium micranthum
arenaria Bentham	G. tenuiflora
arizonica (Greene) Rydberg	Ipomopsis aggregata
aristella Gray	Collomia tinctoria
ashtonae Nelson	Linanthus aureus
atractyloides Steudel	Navarretia atractyloides
atrata Jones	Navarretia divaricata
attenuata (Gray) Nelson	Ipomopsis aggregata
aurea Nuttall	Linanthus aureus
australis Heller ms.	Ipomopsis aggregata
australis Grant & Grant	12–v
bakeri Greene ms	G. subnuda
bella Gray	Linanthus bellus
berterii De Candolle	Navarretia involucrata
beyrichiana Bouché	Ipomopsis rubra
bicolor (Nuttall) Brand	Linanthus bicolor
biflora (Ruiz & Pavon) Macbride	Collomia biflora
bigelovii Gray	Linanthus bigelovii
bolanderi Gray	Linanthus bolanderi
brachysiphon Wooton & Standley	Ipomopsis polyantha
brandegei Gray	Polemonium brandegei
brauntonii Jepson & Mason	Eriastrum puriflorum
brecciarum Jones	12–iv
brevicula Gray	Linanthus breviculus
breweri Gray	Navarretia breweri
bridgesii (Gray) Wherry	Ipomopsis aggregata
burleyana Nelson	Ipomopsis congesta
caespitosa Gray	acc. Brand not Polemoniaceae
caespitosa Nelson	Leptodactylon caespitosum
calcarea Jones	G. pinnatifida
californica Hooker & Arnott	Leptodactylon californicum
calothyrsa Johnston	near Ipomopsis macombii
calyptrata Rivers	Anthodiscus peruanus (Caryocaraceae)
campanulata Gray	12–i
campylantha Wooton & Standley	near Ipomopsis multiflora
cana (Jones) Heller	12–iv
candida Rydberg	Ipomopsis aggregata
capillare Kellogg	see G. capillaris
capillaris Kellogg	12–v
capitata Sims	12–iii
caruifolia Abrams	12–v
cephaloidea Rydberg	Ipomopsis spicata
chachanensis Johnston	G. glutinosa Philippi
chamissonis Greene	G. capitata
ciliata Bentham	Linanthus ciliatus
clivorum (Jepson) Grant	12–iii
clokeyi Mason	G. ophthalmoides
cobijanensis Brand	G. glutinosa Philippi

...(*continued*)

Gilia

collina Eastwood	G. leptantha
columbiana Piper ex Brand	G. capillaris
congesta Hooker	Ipomopsis congesta
copiapina Philippi	G. crassifolia
coronopifolia Persoon	Ipomopsis rubra
cotulaefolia Steudel	Navarretia cotulaefolia
crandallii Rydberg	G. subnuda
crassifolia Bentham	12–iv
crebrifolia Nuttall	Ipomopsis congesta
dactylophyllum Torrey	Linanthus demissus
davyi Milliken	G. latiflora
debilis Watson	Collomia debilis
demissa Gray	Linanthus demissus
densiflora Bentham	Linanthus grandiflorus
densifolia Bentham	Eriastrum densifolium
depressa Jones ex Gray	Ipomopsis depressa
diegensis (Munz) Grant & Grant	12–iv
dianthiflora Steudel	Linanthus dianthiflorus
dianthoides Endlicher	Linanthus dianthiflorus
dichotoma Bentham	Linanthus dichotomus
diffusa Congdon	G. tricolor
diffusa Philippi	Polemonium micranthum
dissecta Heller ex Brand	G. capitata
divaricata Nuttall	Allophyllum divaricatum
divaricata Torrey	Navarretia divaricata
douglasii Gillies ex Bentham	Ipomopsis gossypifera
dunnii Kellogg	G. effusa
eastwoodiae Brand	Linanthus serrulatus
eastwoodiae Heller	Linanthus serrulatus
effusa (Gray) Macbride	12–i
elmeri Piper ms	Collomia tinctoria
elongata Steudel	Eriastrum densifolium
erecta Hieronymus	G. laciniata
eremica Craig	Eriastrum eremicum
eryngoides Lehmann	Navarretia involucrata
exigua Brand	Linanthus bicolor
exilis Abrams	G. ochroleuca
exserta Nelson	Ipomopsis polyantha
fenzlia Steudel	Linanthus dianthiflorus
filicaulis Torrey & Gray	Navarretia filicaulis
filifolia Nuttall	Eriastrum filifolium
filiformis Parry	12–i
filipes Bentham	Linanthus filipes
flavocincta Nelson	G. ophthalmoides
floccosa Gray	Eriastrum luteum
floribunda Gray	Linanthus nuttallii
floridana Don	Ipomopsis rubra
foetida Gillies ex Bentham	12–i
formosa Greene ex Brand	near G. subnuda
formosissima Wooton & Standley	Ipomopsis aggregata
frutescens Rydberg	Ipomopsis frutescens

...(*continued*)

Gilia

gayana Weddell	Polemonium micranthum
gilioides (Bentham) Greene	Allophyllum gilioides
gilmanii Jepson	G. ripleyi
glabrata Philippi	G. glutinosa Philippi
glandulifera Heller	G. capitata
glandulosa Scheele	G. rigidula
globularis Brand	Ipomopsis spicata
glomeriflora (Jussieu) Bentham	Ipomopsis pinnata
gloriosa Brandegee	Ipomopsis gloriosa
glutinosa (Bentham) Gray	Allophyllum glutinosum
glutinosa Philippi	12–i
gossypifera Gillies ex Bentham	Ipomopsis gossypifera
gracilis (Douglas) Hooker	Microsteris gracilis
gracilis Philippi	Salpiglossis parviflora Philippi (Solana-
	ceae)
graciosa (Milliken) Brand	Linanthus androsaceus
grandiflora Gray	Collomia grandiflora
grandiflora Steudel	Linanthus grandiflorus
grayi Nelson	acc. Brand not Polemoniaceae
greeneana Wooton & Standley	Ipomopsis aggregata
grinnellii Brand	G. splendens
guadalupensis Brand	Linanthus pygmaeus
gunnisonii Torrey & Gray	Ipomopsis gunnisonii
guttata Gray	Ipomopsis tenuifolia
hallii Parish	Leptodactylon pungens
hamata (Greene) Munz	Navarretia hamata
harknessii Curran	Linanthus harknessii
havardi Gray	Ipomopsis havardi
haydeni Gray	G. subnuda
heterodoxa Greene	Navarretia heterodoxa
heterophylla Douglas ex Hooker	Collomia heterophylla
hispida Piper	acc. Piper and Brand not Polemoniaceae
hoffmanni Eastwood	G. tenuiflora
hookeri Bentham	Leptodactylon pungens
howardi Jones	Collomia debilis
hugelia Steudel	Eriastrum densifolium
humilis (Greene) Piper	Microsteris gracilis
hutchinsifolia Rydberg	12–ii
hybrida (Vilmorin) Hort.	Linanthus androsaceus
iberidifolia Bentham	Ipomopsis congesta
incisa Bentham	12–i
inconspicua Douglas ex Hooker	G. inconspicua (Smith) Sweet
inconspicua (Smith) Sweet	12–iv
insignis (Brand) Cory & Parks	G. rigidula
intermedia Philippi	G. crassifolia
interior (Mason & Grant) Grant	12–iv
intertexta Steudel	Navarretia intertexta
involucrata Philippi	Ipomopsis gossypifera
inyoensis Johnston	12–i
jaegeri Munz	Leptodactylon jaegeri
jaredii Schumann	Navarretia mitracarpa
jassajarae Brand	see G. tassajarae

...(*continued*)

Gilia

johowi Meigen	Polemonium micranthum
jokowski Meigen	Polemonium micranthum
jonesii Gray	Linanthus jonesii
kennedyi Porter	Linanthus parryae
klikitatensis Suksdorf	Navarretia tagetina
laciniata Bentham	G. valdiviensis
laciniata Ruiz & Pavon	12–iii
lanata Walpers	Eriastrum sapphirinum
lanigera Philippi	G. crassifolia
lanuginosa Philippi	G. crassifolia
larseni Gray	Collomia debilis
latiflora Gray	12–iv
latifolia Watson	12–i
laxa Vasey & Rose	Linanthus laxus
laxiflora Osterhout	Ipomopsis laxiflora
lemmoni Gray	Linanthus lemmonii
leptalea (Gray) Greene	12–v
leptantha Parish	12–iv
leptomeria Gray	12–ii
leptosiphon Steudel	Linanthus grandiflorus
leptotes Gray	Collomia tenella
leucocephala Gray	Navarretia leucocephala
liebmanni Hort.	G. achilleaefolia
lilacina (Greene) Brand	Leptodactylon pungens
linanthus Steudel	Linanthus dichotomus
lindheimeriana Scheele	G. incisa
linearifolia Howell	G. capillaris
linearis Gray	Collomia linearis
lineata Davidson	G. ochroleuca
liniflora Bentham	Linanthus liniflorus
lithospermoides Brandegee	nomen dubium, type lost
longiflora Philippi ms	G. crassifolia
longiflora (Torrey) Don	Ipomopsis longiflora
longisepala Gandoger	Microsteris gracilis
longituba Bentham	Linanthus androsaceus
lutea (Bentham) Steudel	Linanthus luteus
lutescens Steudel	Eriastrum luteum
macgregorii (Brand) Mason, nomen nudum	Navarretia peninsularis
macombii Torrey ex Gray	Ipomopsis macombii
maculata Parish	Linanthus maculatus
mariposiana (Milliken) Brand	Linanthus serrulatus
matthewsii Gray	Langloisia matthewsii
mazama (Coville) Nelson & Macbride	Collomia mazama
mcvickerae Jones	12–ii
mellita Greene	Navarretia mellita
merrillii Nelson	Ipomopsis congesta
mexicana Grant & Grant	12–iv
micrantha Gray	Linanthus androsaceus
micrantha Steudel	Linanthus androsaceus
micromeria Gray	12–ii

...(*continued*)

Gilia

microsteris Piper	Microsteris gracilis
millefolia Brand ms	G. nevinii
millefoliata Fischer & Meyer	12–iii
minima Gray	Navarretia minima
minor Grant & Grant	12–iv
minutiflora Bentham	Ipomopsis minutiflora
modesta Hall	Linanthus concinnus
modesta Philippi	G. crassifolia
modocensis Eastwood	12–iv
mohavensis (Mason) Putnam	Linanthus arenicola
montana Nelson & Kennedy	Ipomopsis congesta
montana (Greene) Parish	Linanthus montanus
montezumae Tidestrom & Dayton	G. subnuda
mucronata Lehmann	Navarretia involucrata
multicaulis Bentham	G. achilleaefolia
multiflora Nuttall	Ipomopsis multiflora
navarretia Steudel	Navarretia involucrata
nevadensis Tidestrom	Ipomopsis congesta
nevinii Gray	12–iii
nivalis Hérincq	G. tricolor
nuda Rydberg	Ipomopsis congesta
nudata (Greene) Brand	Linanthus nudatus
nudicaulis Gray	Gymnosteris nudicaulis
nudicaulis Philippi	may not belong to Polemoniaceae
nuttallii Gray	Linanthus nuttallii
oblanceolata (Eastwood) Brand	Linanthus oblanceolatus
ocellata Schumann	Navarretia nigellaeformis
ochroleuca Jones	12–iv
ophthalmoides Brand	12–iv
orcuttii Parry & Gray	Linanthus orcuttii
oreophila Greene	G. achilleaefolia
pacifica (Milliken) Brand	Linanthus orcuttii
pallida Heller	G. capitata
palmeri Watson.	12–i
palmifrons (Brand) Rydberg	Ipomopsis congesta
parishii Peter ms	G. achilleaefolia
parryae Gray	Linanthus parryae
parviflora Sprengel	Gilia inconspicua
parvula Greene	Navarretia heterodoxa
parvula Rydberg	Gymnosteris nudicaulis
patagonica Spegazzini	near G. crassifolia
peduncularis Eastwood ex Milliken	G. achilleaefolia
pedunculata Eastwood	G. achilleaefolia
peninsularis (Greene) Munz	Navarretia peninsularis
pentstemonoides Jones	12–ii
pharnaceoides Bentham	Linanthus liniflorus
pinnata (Cavanilles) Brand	Ipomopsis pinnata
pinnatifida Nuttall	12–ii
pinnatifida Sessé & Mociño ex Don	Collomia heterophylla
platyloba Johnston	G. rigidula
pluriflora Heller	Eriastrum pluriflorum
polyantha Rydberg	Ipomopsis polyantha

...(*continued*)

Gilia

polycladon Torrey   Ipomopsis polycladon
pringlei Gray   near Ipomopsis macombii
prostrata Gray   Navarretia prostrata
pubescens Steudel   Navarretia pubescens
pulchella Douglas ex Hooker   Ipomopsis aggregata

pumila Nuttall   Ipomopsis pumila
pungens Hooker   Navarretia squarrosa
punctata (Gray) Munz   Langloisia punctata
pungens (Torrey) Bentham   Leptodactylon pungens
purpusii Brandegee   12–i

pusilla Bentham   Linanthus pusillus
pygmaea Brand   Linanthus pymaeus

ramosissima Philippi   G. glutinosa Philippi
rattanii Gray   Linanthus rattanii
rawsoniana (Greene) Macbride   Collomia rawsoniana
rigidula Bentham   12–i
ripleyi Barneby   12–i

roseata Rydberg   Ipomopsis roseata
royalis Brand   Linanthus breviculus
rubra (L.) Heller   Ipomopsis rubra

salticola Eastwood   G. leptantha
sapphirina Eastwood   Eriastrum sapphirinum
scabra Brandegee   Linanthus nuttallii
scariosa Rydberg   Ipomopsis aggregata
schotti Watson   Langloisia schotti

scopulorum Jones   12–v
sedifolia Brandegee   near G. stenothyrsa
sessei Don   Collomia heterophylla
setosissima Gray   Langloisia setosissima
sherman-hoytae Craig   Eriastrum pluriflorum

sinister Jones   see G. sinistra
sinistra Jones   G. capillaris
sinuata Douglas   12–iv
sonorae Rose   Ipomopsis sonorae
sparsiflora Eastwood   Eriastrum sparsiflorum

spergulifolia Rydberg   Ipomopsis congesta
spicata Nuttall   Ipomopsis spicata
splendens Douglas   12–v
squarrosa Hooker & Arnott   Navarretia squarrosa
staminea Greene   G. capitata

stellata Heller   12–v
stenothyrsa Gray   12–ii
stewartii Johnston   G. rigidula
straminea Rydberg   G. sinuata
stricta Scheele   nomen dubium near G. achilleaefolia

subacaulis Rydberg   G. leptomeria
subalpina Greene ex Brand   G. capillaris
subnuda Gray   12–ii
superba Eastwood   G. subnuda

tassajarae Brand   Linanthus androsaceus
tenella Bentham   Linanthus bicolor
tenella Nuttall ex Gray   Linanthus liniflorus

... (continued)

Gilia

tenerrima Gray	12–v
tenuiflora Bentham	12–iv
tenuifolia Gray	Ipomopsis tenuifolia
tenuiloba Parish	Leptodactylon pungens
tenuisecta Heller	G. capitata
tenuituba Rydberg	Ipomopsis tenuituba
tetrabreccia Grant & Grant	G. modocensis
texana Wooton & Standley	Ipomopsis aggregata
thurberi Torrey ex Gray	Ipomopsis thurberi
tinctoria Kellogg	Collomia tinctoria
tomentosa Martens & Galeotti	Ipomopsis pinnata
transmontana Grant	12–iv
traskiae Eastwood ex Milliken	Allophyllum glutinosum
tricolor Bentham	12–iii
tridactyla Rydberg	Ipomopsis spicata
trifida Bentham	Ipomopsis pumila
trifida Nuttall	Ipomopsis spicata
triodon Eastwood	G. leptomeria
truncata Davidson	Ipomopsis tenuifolia
tularensis Brand	Linanthus oblanceolatus
tweedyi Rydberg	12–iv
uncialis Brandegee	12–i
valdiviensis Grisebach	12–iii
veatchii Parry	Leptodactylon veatchii
violacea Heller	Allophyllum violaceum
virgata Steudel	Eriastrum virgatum
viscida Wooton & Standley	Gilia pinnatifida
viscidula Gray	Navarretia viscidula
watsoni Gray	Leptodactylon watsonii
wilcoxii Nelson	Eriastrum wilcoxii
wrightii Gray	near Ipomopsis frutescens

Giliastrum

acerosum (Gray) Rydberg	Gilia rigidula
rigidulum (Bentham) Rydberg	Gilia rigidula

Gymnosteris

leibergii Brand	G. parvula
minuscula Jepson	G. parvula
nudicaulis (Hooker & Arnott) Greene	9
parvula (Rydberg) Heller	9
pulchella Greene	G. nudicaulis
rydbergii Tidestrom	G. parvula

Hoitzia

amplectens Hooker & Arnott	Loeselia amplectens
aristata Humboldt, Bonpland & Knuth	Loeselia ciliata
caerulea Cavanilles	Loeselia coerulea
capitata Willdenow	Loeselia glandulosa
cervantesii Humboldt. Bonpland & Knuth	Loeselia glandulosa

... *(continued)*

Hoitzia

coccinea Cavanilles	Loeselia mexicana
coerulea Cavanilles	Loeselia coerulea
conglomerata Humboldt, Bonpland & Knuth	Loeselia glandulosa
elata Hooker & Arnott	Loeselia glandulosa
floribunda Martens & Galeotti	Loeselia coerulea
glandulosa Cavanilles	Loeselia glandulosa
linearis Sprengel	Collomia linearis Nuttall
loeselia Sprengel	Loeselia ciliata
lupulina Hooker & Arnott	Loeselia ciliata
mexicana Lamarck	Loeselia mexicana
nepetaefolia Chamisso & Schlechtendahl	Loeselia glandulosa
pumila Martens & Galeotti	Loeselia pumila
ramosissima Martens & Galeotti	Loeselia glandulosa
scabra Martens & Galeotti	Loeselia glandulosa
scariosa Martens & Galeotti	Loeselia scariosa
spicata Willdenow ex Roemer & Schultes	Loeselia glandulosa
squarrosa Eschscholtz	Navarretia squarrosa

Hugelia

miscellaneous species	Trachymene (Umbelliferae)
abramsii Jepson & Bailey	Eriastrum abramsii
brauntonii Jepson	Eriastrum pluriflorum
densifolia Bentham	Eriastrum densifolium
diffusa (Gray) Jepson	Eriastrum diffusum
elongata Bentham	Eriastrum densifolium
eremica Jepson	Eriastrum eremicum
filifolia Jepson	Eriastrum filifolium
floccosa Nuttall	Eriastrum luteum
hooveri Jepson	Eriastrum hooveri
lanata Lindley	Eriastrum pluriflorum
lutea Bentham	Eriastrum luteum
pluriflora Ewan	Eriastrum pluriflorum
virgata Bentham	Eriastrum virgatum

Huthia

coerulea Brand	3
longiflora Brand	3

Ipomeria

aggregata (Pursh) Nuttall.	Ipomopsis aggregata
albida Nuttall	Gilia laciniata
coronopifolia (Willdenow) Nuttall	Ipomopsis rubra
inconspicua (Smith) Nuttall	Gilia inconspicua

Ipomoea

miscellaneous species	Ipomoea (Convolvulaceae)
rubra Murray	Ipomopsis rubra

Ipomopsis

aggregata (Pursh) Grant	13–ii
congesta (Hooker) Grant	13–iii
depressa (Jones) Grant	13–iii

...(*continued*)

Ipomopsis

elegans Lindley	I. aggregata
elegans Michaux	I. rubra
frutescens (Rydberg) Grant	13–iii
gloriosa (Brandegee) Grant	13–i
gossypifera (Gillies) Grant	13–iii
gunnisonii (Torrey & Gray) Grant	13–iii
havardi (Gray) Grant	13–i
inconspicua Smith	Gilia inconspicua
laxiflora (Coulter) Grant	13–ii
longiflora (Torrey) Grant	13–ii
macombii (Torrey) Grant	13–ii
minutiflora (Bentham) Grant	13–iii
multiflora (Nuttall) Grant	13–i
picta Paxton	I. rubra
pinnata (Cavanilles) Grant	13–i
polyantha (Rydberg) Grant	13–i
polycladon (Torrey) Grant	13–iii
pumila (Nuttall) Grant	13–iii
roseata (Rydberg) Grant	13–iii
rubra (Linnaeus) Wherry	13–ii
sonorae (Rose) Grant	13–iii
spicata (Nuttall) Grant	13–iii
tenuifolia (Gray) Grant	13–i
tenuituba (Rydberg) Grant	13–ii
thurberi (Torrey) Grant	13–ii

Langloisia

flaviflora Davidson	L. schottii
lanata Brand	15
matthewsii (Gray) Greene	15
punctata (Coville) Goodding	15
schottii (Torrey) Greene	15
setosissima (Torrey & Gray) Greene	15

Leptodactylon

brevifolium Rydberg	L. pungens
caespitosum Nuttall	17
californicum Hooker & Arnott	17
floribundum (Gray) Rydberg	Linanthus nuttallii
gloriosum (Brandegee) Wherry	Ipomopsis gloriosa
hallii (Parish) Heller	L. pungens
hazeliae Peck	L. pungens
jaegeri (Munz) Wherry	17
lilacinum Greene ex Baker, nomen nudum	L. pungens
melingii Wiggins	Linanthus melingii
nuttallii (Gray) Rydberg	Linanthus nuttallii
patens Heller	L. pungens
pungens (Torrey) Rydberg	17
tenuilobum (Parish) Heller	L. pungens
veatchii (Parry ex Greene) Wherry	17
watsonii (Gray) Rydberg	17

Leptosiphon

acicularis (Greene) Jepson	Linanthus acicularis
androsaceus Bentham	Linanthus androsaceus

...(*continued*)

Leptotosiphon

aureus Bentham	Linanthus aureus
bicolor Nuttall	Linanthus bicolor
ciliatus (Bentham) Jepson	Linanthus ciliatus
densiflorus Bentham	Linanthus grandiflorus
grandiflorus Bentham	Linanthus grandiflorus
hybridus Vilmorin	Linanthus androsaceus
luteus Bentham	Linanthus androsaceus
mariposianus Heller	Linanthus serrulatus
parviflorus Bentham	Linanthus androsaceus
roseus Thompson	Linanthus androsaceus

Linanthastrum

floribundum (Gray) Wherry	Linanthus nuttallii
nuttallii (Gray) Ewan	Linanthus nuttallii
melingii (Wiggins) Wherry	Linanthus melingii

Linanthus

acicularis Greene	18–iii
ambiguus (Rattan) Greene	18–iv
androsaceus (Bentham) Greene	18–iii
arenicola (Jones) Jepson & Bailey	18–vi
asprellus Greene	L. bicolor
aureus (Nuttall) Greene	18–iv
bakeri Mason	18–iv
bellus (Gray) Greene	18–vi
bicolor (Nuttall) Greene	18–iii
bigelovii (Gray) Greene	18–v
bolanderi (Gray) Greene	18–iv
breviculus (Gray) Greene	18–iii
ciliatus (Bentham) Greene	18–iii
concinnus Milliken	18–v.
croceus Eastwood	L. androsaceus
dactylophyllum Rydberg	L. demissus
demissum Mason	see L. demissus
demissus (Gray) Greene	18–vi
densiflorus (Bentham) Jepson	L. grandiflorus
dianthiflorus (Bentham) Greene	18–vi
dichotomus Bentham	18–v
diffusus Heller ms.	L. bicolor
eastwoodiae Heller	L. bicolor
filipes (Bentham) Greene	18–iv
floribundus Greene	L. nuttallii
graciosus Milliken	L.androsaceus
grandiflorus (Bentham) Greene	18–iii
harknessii (Curran) Greene	18–iv
jonesii (Gray) Greene	18–v
killipii Mason	18–vi
laxus (Vasey & Rose) Wherry	18–i
lemmonii (Gray) Greene	18–iv
liniflorus (Bentham) Greene	18–iv
longitubus Heller	L. androsaceus
luteolus Greene	L. androsaceus
maculatus (Parish) Milliken	18–vi
mariposianus Milliken	L. serrulatus
melingii (Wiggins) Grant	18–i

...(*continued*)

Linanthus

mohavensis Mason	L. arenicola
montanus Greene	18–iii
nashianus Jepson	L. nudatus
neglectus Greene	L. ciliatus
nudatus Greene	18–iii
nudicaulis (Hooker & Arnott) Howell	Gymnosteris nudicaulis
nuttallii (Gray) Greene	18–i
oblanceolatus (Eastwood) Brand	18–iii
ocellatus Heller	L. bicolor
orcuttii (Parry & Gray) Jepson	18–vi
pacificus Milliken	L. orcuttii
parryae (Gray) Greene	18–vi
parviflorus (Bentham) Greene	L. androsaceus
peirsonii Mason	L. bellus
pharnaceoides (Bentham) Greene	L. liniflorus
plaskettii Eastwood	L. androsaceus
pusillus (Bentham) Greene	18–iv
pygmaeus (Brand) Howell	18–iv
rattanii (Gray) Greene	18–iv
rosaceus Greene	L. androsaceus
saxiphilus Davidson	L. nuttallii
septentrionalis Mason	18–iv
serrulatus Greene	18–iii
tularensis (Brand) Mason	L. oblanceolatus
wigginsii Mason	L. laxus

Loeselia

amplectens (Hooker & Arnott) Bentham	5–i
aristata Don	L. ciliata
carionis Peter	acc. Brand possibly Scrophulariaceae
cervantesii Don	L. glandulosa
ciliata Linnaeus	5–i
coccinea Don	L. mexicana
coerulea (Cavanilles) Don	5–i
columbiana Gandoger	L. glandulosa
conglomerata Don	L. glandulosa
cordifolia Hemsley & Rose	L. amplectens
effusa Gray	Gilia effusa
glandulosa (Cavanilles) Don	5–i
gloriosa (Brandegee) Johnston	Ipomopsis gloriosa
grandiflora Standley	5–ii
greggii Watson	L. scariosa
guttata Gray	Ipomopsis tenuifolia
havardi Gray	Ipomopsis havardi
intermedia Loesener	L. pumila
involucrata Don	5–i
matthewsii Gray	Langloisia matthewsii
mexicana (Lamarck) Brand	5–i
nepetaefolia Don	L. glandulosa
pumila (Martens & Galeotti) Walpers	5–i
purpusii Brandegee	Gilia purpusii
ramosissima Walpers	L. glandulosa
rupestris Bentham	L. coerulea
scabra Walpers	L. glandulosa
scariosa (Martens & Galeotti) Walpers	5–i

...(*continued*)

Loeselia
schottii Gray
setosissima Gray
tenuifolia Gray

Langloisia schottii
Langloisia setosissima
Ipomopsis tenuifolia

Lychnidea
miscellaneous species

Lychnidea Burm. (Manulea, Scrophulariaceae)

blattariae Plukenet
caroliniana Martyn
fistulosa Plukenet
flore purpureo Lawson

Phlox nivalis
Phlox ovata
Phlox ovata
Phlox carolina

folio melampyri Dillenius
folio salicino Dillenius
mariana elatior Plukenet
marilandica Rajus
sempervirens Collinson

Phlox glaberrima
Phlox paniculata
Phlox maculata
Phlox pilosa L.
Phlox subulata

umbellifera Plukenet
virginiana blattariae Plukenet

Phlox pilosa L.
Phlox paniculata

Lychnoides
marilandica Ray

Phlox pilosa L.

Microsteris
andicola (Bentham) Greene

nomen dubium, Collomia, Microsteris, or Gilia?

californica Greene
depressa Davidson & Moxley
diffusa Heller
gilioides Davidson & Moxley

M. gracilis
Ipomopsis depressa
M. gracilis
Allophyllum gilioides

glabella Greene
gracilis (Hooker) Greene
heterophylla Brockman
humilis (Douglas) Greene
larseni Brockman
macdougalii Heller
micrantha Greene

M. gracilis
11
Collomia heterophylla
M. gracilis
Collomia larseni
M. gracilis
M. gracilis

stricta Greene
traskiae Davidson & Moxley

M. gracilis
Allophyllum glutinosum

Myotoca
eritrichioides Grisebach ms

Microsteris gracilis

Navarretia
abramsii Elmer
achilleaefolia (Bentham) Kuntze
aggregata (Pursh) Kuntze
andicola (Philippi) Kuntze
androsacea (Bentham) Kuntze

Eriastrum abramsii
Gilia achilleaefolia
Ipomopsis aggregata
Gilia crassifolia
Linanthus androsaceus

aristella (Gray) Kuntze
atractyloides (Bentham) Hooker & Arnott
aurea (Nuttall) Kuntze

Collomia tinctoria
16–i
Linanthus aureus

bakeri Mason
bella (Gray) Kuntze

16–iv
Linanthus bellus

biflora (Ruiz & Pavon) Kuntze

Collomia biflora

...(*continued*)

Navarretia

bolanderi (Gray) Kuntze	Linanthus bolanderi
bowmanae Eastwood	N. cotulaefolia
brandegei (Gray) Kuntze	Polemonium brandegei
brevicula (Gray) Kuntze	Linanthus breviculus
breweri (Gray) Greene	16–ii
caespitosa (Gray) Kuntze	Gilia caespitosa, q.v.
californica (Hooker & Arnott) Kuntze	Leptodactylon californicum
campanulata (Gray) Kuntze	Gilia campanulata
capillaris (Kellogg) Kuntze	Gilia capillaris
capitata (Sims) Kuntze	Gilia capitata
ciliata (Bentham) Kuntze	Linanthus ciliatus
congesta (Hooker) Kuntze	Ipomopsis congesta
cotulaefolia (Bentham) Hooker & Arnott	16–iii
crassifolia (Bentham) Kuntze	Gilia crassifolia
debilis (Watson) Kuntze	Collomia debilis
demissa (Gray) Kuntze	Linanthus demissus
densiflora (Bentham) Kuntze	Linanthus grandiflorus
densifolia (Bentham) Kuntze	Eriastrum densifolia
depressa (Jones) Kuntze	Ipomopsis depressa
dianthiflora (Bentham) Kuntze	Linanthus dianthiflorus
dichotoma (Bentham) Kuntze	Linanthus dichotomus
diffusa (Philippi) Kuntze	Polemonium micranthum
divaricata (Torrey) Greene	16–ii
diversifolia (Greene) Kuntze	Collomia diversifolia
dubia Brand	N. filicaulis
eastwoodiae Brand	N. mellita
effusa (Gray) Kuntze	Gilia effusa
erecta Heller	N. tagetina
eriocephala Mason	16–iii
eritrichioides (Grisebach) Kuntze	Microsteris gracilis
eritrichodes Kuntze	see N. eritrichioides
erythraeodes Kuntze	see N. erythraeoides
erythraeoides (Grisebach) Kuntze	Microsteris gracilis
fallax Brand	N. heterodoxa
filicaulis (Torrey) Greene	16–i
filifolia (Nuttall) Kuntze	Eriastrum filifolium
filiformis (Parry) Kuntze	Gilia filiformis
floccosa (Gray) Kuntze	Eriastrum luteum
floribunda (Gray) Kuntze	Linanthus nuttallii
foetida (Gillies) Kuntze	Gilia foetida
foliacea Greene	N. hamata
gayana (Weddell) Kuntze	Polemonium micranthum
giliodes Kuntze	see N. gilioides
gilioides (Bentham) Kuntze	Allophyllum gilioides
glutinosa (Presl) Kuntze	Cantua glutinosa, q.v.
gossypifera (Gillies) Kuntze	Ipomopsis gossypifera
gracilis (Douglas) Kuntze	Microsteris gracilis
grandiflora (Douglas) Kuntze	Collomia grandiflora
gunnisonii (Torrey & Gray) Kuntze	Ipomopsis gunnisonii
guttata (Gray) Kuntze	Ipomopsis tenuifolia
hamata Greene	16–i

...(*continued*)

Navarretia

harcknessii Kuntze	see N. harknessii
harknessii (Curran) Kuntze	Linanthus harknessii
havardii (Gray) Kuntze	Ipomopsis havardi
haydenii (Gray) Kuntze	G. subnuda
helleri Brand	N. squarrosa x N. mellita
heterandra Mason	16–iii
heterodoxa Greene	16–i
heterophylla Bentham	Collomia heterophylla
hirsutissima Brand	N. atractyloides
iberidifolia (Bentham) Smyth	Ipomopsis congesta
incisa (Bentham) Kuntze	Gilia incisa
inconspicua (Smith) Kuntze	Gilia inconspicua
intertexta (Bentham) Hooker	16–iii
involucrata Ruiz & Pavon	16–iv
jaredii Eastwood	N. mitracarpa
jepsonii Bailey	16–iii
jonesii (Gray) Kuntze	Linanthus jonesii
klickitatensis Suksdorf	see N. klikitatensis
klikitatensis Suksdorf	N. tagetina
laciniata (Ruiz & Pavon) Kuntze	Gilia laciniata
lanuginosa (Philippi) Kuntze	Gilia crassifolia
latiflora (Gray) Kuntze	Gilia latiflora
latifolia (Watson) Kuntze	Gilia latifolia
lemmonii (Gray) Kuntze	Linanthus lemmonii
leptantha Greene	N. hamata
leptomeria (Gray) Kuntze	Gilia leptomeria
leptotes (Gray) Kuntze	Collomia tenella
leucocephala Bentham	16–iv
linearis (Nuttall) Kuntze	Collomia linearis Nuttall
liniflora (Bentham) Kuntze	Linanthus liniflorus
longiflora (Torrey) Kuntze	Ipomopsis longiflora
longifolia Rydberg	see N. longiflora
lutea (Bentham) Kuntze	Linanthus luteus
lutea Brand	Eriastrum luteum
lutescens (Steudel) Kuntze	Eriastrum luteum
macgregorii Brand	N. peninsularis
macombii (Torrey) Kuntze	Ipomopsis macombii
macrantha Brand	N. hamata
matthewsii (Gray) Kuntze	Langloisia matthewsii
mellita Greene	16–i
micromeria (Gray) Kuntze	Gilia micromeria
millefolia (Gray) Kuntze	Gilia nevinii
minima Nuttall	16–iv
minutiflora (Bentham) Kuntze	Ipomopsis minutiflora
mitracarpa Greene	16–iii
multicaulis (Bentham) Kuntze	Gilia achilleaefolia
multiflora (Nuttall) Kuntze	Ipomopsis multiflora
nigellaeformis Greene	16–iii
nudicaulis (Hooker & Arnott) Kuntze	Gymnosteris nudicaulis
nuttallii (Gray) Kuntze	Linanthus nuttallii

... *(continued)*

Navarretia

ocellata Eastwood	N. nigellaeformis
orcuttii (Parry) Kuntze	Linanthus orcuttii
parryae (Gray) Kuntze	Linanthus parryae
parviflora (Torrey) Kuntze	Linanthus bigelovii
parvula Greene	N. heterodoxa
pauciflora Mason	16–iv
peninsularis Greene	16–ii
philippiana Kuntze	Gilia glutinosa Phil.
pilosifaucis St. John & Weitman	N. intertexta
pinnata (Cavanilles) Kuntze	Ipomopsis pinnata
pinnatifida (Nuttall) Kuntze	Gilia pinnatifida
pleiantha Mason	16–iv
plieantha Mason	see N. pleiantha
polycladon (Torrey) Kuntze	Ipomopsis polycladon
prolifera Greene	16–ii
propinqua Suksdorf	16–iii
prostrata (Gray) Greene	16–iv
pterosperma Eastwood	N. squarrosa
pubescens (Bentham) Hooker & Arnott	16–iii
pumila (Nuttall) Kuntze	Ipomopsis pumila
pungens Hooker	N. squarrosa
pungens (Torrey) Kuntze	Leptodactylon pungens
purpurea Greene ex Brand	N. viscidula
pusilla (Bentham) Kuntze	Linanthus pusillus
pusilla Hort. ex Steudel	N. involucrata
rattanii (Gray) Kuntze	Linanthus rattanii
rigidula (Bentham) Kuntze	Gilia rigidula
rosulata Brand	N. heterodoxa
rubra (Linnaeus) Kuntze	Ipomopsis rubra
savagei Henderson	N. tagetina
schottii Torrey	Langloisia Schottii
setiloba Coville	16–iii
setosissima Torrey & Gray	Langloisia setosissima
spicata (Nuttall) Kuntze	Ipomopsis spicata
squarrosa (Eschscholtz) Hooker & Arnott	16–i
stenosiphon Kuntze	Collomia cavanillesii
stenothyrsa (Gray) Kuntze	Gilia stenothyrsa
stricta Howell	N. intertexta
subnuda (Torrey) Kuntze	Gilia subnuda
subuligera Greene	16–iv
suksdorfii Howell	N. minima
tagetina Greene	16–iii
tenerrima (Gray) Kuntze	Gilia tenerrima
tenuiflora (Bentham) Kuntze	Gilia tenuiflora
tenuifolia (Gray) Kuntze	Ipomopsis tenuifolia
thurberi (Gray) Kuntze	Ipomopsis thurberi
tricolor (Bentham) Kuntze	Gilia tricolor
valdiviensis (Grisebach) Kuntze	Gilia valdiviensis
virgata (Bentham) Kuntze	Eriastrum virgatum
viscidula Bentham	16–ii

...(*continued*)

Navarretia

watsonii (Gray) Kuntze	Leptodactylon watsonii
wilcoxii (Nelson) Brand	Eriastrum wilcoxii
wrightii (Gray) Brand	near Ipomopsis frutescens

Periphragmos

corymbosa Ruiz ms.	Cantua pyrifolia
dependens Ruiz & Pavon	Cantua buxifolia
flexuosus Ruiz & Pavon	Cantua pyrifolia
uniflorus Ruiz & Pavon	Cantua buxifolia

Phacelia

miscellaneous species	Phacelia (Hydrophyllaceae)
furcata Douglas ex Hooker	Ipomopsis congesta

Phlox

abdita Nelson	P. alyssifolia
acerba Nelson	P. austromontana
aciculifolia Kennedy	P. austromontana
aculeata Nelson	10–iii
acuminata Pursh	P. paniculata
acutifolia Sweet, nomen nudum	P. paniculata
adsurgens Torrey ex Gray	10–i
alaskensis Jordal	P. richardsoni
alba Moench	P. maculata
albomarginata Jones	10–iii
aleardii Taylor, nomen nudum	P. paniculata
altissima Moench	P. carolina
alyssifolia Greene	10–iii
amabilis Brand	10–i
americana Hort. ex Sweet	P. paniculata
amoena Hort.	P. pilosa × P. subulata
amoena Sims	10–ii
amplexicaulis Rafinesque	nomen dubium, near P. pilosa
amplifolia Britton	10–i
andersonii Rydberg, nomen nudum	near P. albomarginata
andichood Hort.	P. andicola × P. hoodii
andicola Nuttall ex Gray	10–iii
arendsii Hort.	P. divaricata × P. paniculata
argillacea Clute & Ferriss	P. pilosa
aristata Michaux	P. pilosa
aspera Nelson	P. pilosa
assurgens Farrer	see P. adsurgens
atkinsii Taylor, nomen nudum	P. paniculata
atrocaulis Pope ex Loudon	P. paniculata
auriculata Sessé & Mociño	Bonplandia geminiflora
austromontana Coville	10–iii
autumnale Young	nomen dubium
bernardina Munz & Johnston	P. dolichantha
bifida Beck	10–i
biflora Ruiz & Pavon	Collomia biflora
bimaculata Sweet	P. maculata
borealis Wherry	10–iii
brevifolia Baumann ex Hoffmansegg, nomen nudum	P. paniculata

...(*continued*)

Phlox

brevistylis Nelson	P. nana
bridgesii Marnock	nomen dubium, near P. glaberrima
brittonii Small	P. subulata
broughtonii Marnock	nomen dubium, near P. glaberrima
brownii Salm-Dyck, nomen nudum	P. paniculata
bryoides Nuttall	10–iii
buckleyi Wherry	10–i
caesia Eastwood	P. gladiformis
caespitosa Nuttall	10–iii
caldryana Courtois ex Steudel	see P. coldryana
camla Hort.	P. nivalis
camlaensis Hort.	P. nivalis
canadensis Sweet	P. divaricata
canescens Torrey & Gray	P. hoodii
carnea Sims	P. glaberrima
carolina Linnaeus	10–i
carolina Sweet	P. floridana
caroliniana Hill	P. carolina L.
caryophylla Wherry	10–i
cernua Nelson	P. longifolia
cetacea Maund	see P. setacea
ciliata Wehrhahn	P. subulata
clarkioides Poiteau	nomen dubium, near P. glaberrima
cluteana Nelson	10–i
clutena Nelson	see P. cluteana
coldryana Paxton	nomen dubium, near P. glaberrima
collina Rydberg	P. alyssifolia
colubrina Wherry & Constance	10–ii
condensata Nelson	P. caespitosa
cordata Elliott	P. paniculata
cortezana Nelson	P. longifolia
corymbosa Young	P. paniculata
costata Rydberg	P. multiflora
covillei Nelson	10–iii
crassifolia Loddiges	P. stolonifera
criterion Miellez ex Van Houtte	P. paniculata × P. drummondii
cruenta Curtis ex Steudel, nomen nudum	P. paniculata
cultorum Thorsrud & Reisaeter	nomen dubium
cuspidata Scheele	10–ii
cuspidata (Wittmack) Brand	P. drummondii
cyanea Eastwood	P. diffusa
dasyphylla Brand	P. multiflora
decussata Lyon ex Pursh	P. paniculata
dejecta Nelson & Kennedy	P. covillei
densa Brand	P. autromontana
depressa Miellez ex Van Houtte	P. drummondii × P. paniculata
depressa Rydberg	P. multiflora
detonsa Small	P. pilosa
dialypetala Kirschleger	nomen dubium, near P. glaberrima
diapensioides Rydberg	P. albomarginata
diffusa Bentham	10–iii
disticha Sabine ex Young	nomen dubium
divaricata Durand	P. speciosa

...(*continued*)

Phlox

divaricata Linnaeus	10–ii
divaricata Sessé & Mociño	Loeselia sp.
divergens Wenderoth ex Steudel, nomen nudum	P. paniculata
dolichantha Gray	10–i
douglasii Hooker	10–iii
drummondii Hooker	10–ii
elata Penny ex Loudon, nomen nudum	P. paniculata
elegans Taylor, nomen nudum	P. paniculata
elegantissima Taylor, nomen nudum	P. paniculata
excelsa Young	P. paniculata
fimbriata (Wittmack) Brand	P. drummondii
floridana Bentham	10–ii
frondosa Hort	P. subulata × P. nivalis
fruticosa Steudel, nomen nudum	P. paniculata
gilioides Nelson	P. tenuifolia
glaberrima Linnaeus	10–i
glabrata Brand	P. hoodii
glabriflora (Brand) Whitehouse	10–ii
gladiformis (Jones) Nelson	10–iii
glandulosa Shuttleworth ex Gray, nomen subnudum	P. amplifolia
glomerata Nuttall, nomen nudum	P. divaricata
glutinosa Buckley	P. divaricata × P. pilosa
goldsmithii Whitehouse	P. drummondii
gooddingii Nelson & Kennedy	P. gladiformis
gracilis (Hooker) Greene	Microsteris gracilis
grahami Wherry	10–ii
grayi Wooton & Standley	P. stansburyi
griseola Wherry	10–iii
helleri Whitehouse, nomen nudum	P. glabriflora
henryae Wherry	P. nivalis × P. bifida
hentzii Nuttall	P. nivalis
heterophylla Beauvais ex Brand	P. carolina L.
heynholdii Grieve	P. drummondii
hirsuta Nelson	P. stansburyi
hoodii Richardson	10–iii
hookeri Douglas ex Hooker	Leptodactylon pungens
hortorum Bergmans	nomen dubium,
humilis Douglas ms	P. longifolia
hybrida Rafinesque	nomen dubium,
idahonis Wherry	10–i
imminens St. John	P. speciosa
ingramiana Loudon	P. paniculata
insignis De Jonghe	P. paniculata
intermedia Nelson ex Brand	P. multiflora
involucrata Nuttall ex Gray	P. amoena
involucrata Pavon ms	Loeselia ciliata
involucrata Sessé & Mociño	Loeselia involucrata
jenkinsonii Taylor, nomen nudum	P. paniculata
jonesii Wherry	10–iii

...(*continued*)

Phlox

kelseyi Britton   10–iii

laeta Penny ex Loudon, nomen nudum   P. paniculata
lanata Piper   P. hoodii
lanceolata Nelson   P. speciosa
laphami Clute   P. divaricata
latifolia Michaux   P. ovata

leopoldiana Paxton   P. drummondii
lighthipei Small   P. amoena Sims
ligustrifolia Young   nomen dubium
lindheimeri Engelmann ex Gray   P. roemeriana
linearifolia Gray   P. longifolia

linearis Cavanilles   Collomia biflora
listoniana Young   nomen dubium, near P. glaberrima
littoralis Whitehouse   P. glabriflora
longiflora Sweet   P. maculata
longifolia Nuttall   10–i

longituba Heller   P. stansburyi
lorranii Taylor, nomen nudum   P. paniculata

macrantha Brand   P. roemeriana
macrophylla Courtois ex Steudel, nomen   P. paniculata
    nudum
maculata Linnaeus   10–i
maculata Sessé & Mociño   nomen dubium, perhaps Loeselia
marginata (Brand) Nelson   P. longifolia

marianna Lindley ex Vilmorin   P. paniculata
mcallisteri Whitehouse   P. drummondii
melampyrifolia Salisbury   P. glaberrima
mesaleuca Hort. ex Sealy   P. triovulata
mesoleuca Greene   10–ii

mexicana Wherry   10–ii
missoulensis Wherry   10–iii
missourica Salm-Dyck   P. paniculata
mollis Wherry   10–iii
montana Rafinesque   nomen dubium,

multiflora Nelson   10–iii
muscoides Nuttall   P. hoodii

nana Nuttall   10–ii
nelsonii Brand   P. triovulata
nelsonii Hort.   P. subulata
nitida Pursh   near P. ovata
nivalis Hort.   P. subulata

nivalis Sweet   10–ii
nivea Don   see P. nivalis
nuttalliana Don   P. floridana
nuttallii Courtois ex Steudel   P. floridana

obovata Muhlenberg ex Willdenow   P. stolonifera
occidentalis Durand ex Torrey   P. speciosa
odorata Sweet   P. maculata
oklahomensis Wherry   10–ii
oldryana Walpers   see P. coldryana

omniflora Loudon   P. paniculata
ovata Linnaeus   10–i

(continued)

Phlox

paniculata Linnaeus	10–i
patula Nelson	P. multiflora
paxtoni Taylor, nomen nudum	P. paniculata
peckii Wherry	10–iii
pendula Vilmorin	P. paniculata
penduliflora Young	P. paniculata
philadelphica Steudel, nomen nudum	P. paniculata
pilosa Linnaeus	10–ii
pilosa Walter	P. amoena Sims
pinetorum Heller ms	P. speciosa
pinifolia Brand	P. longifolia
pinnata Cavanilles	Ipomopsis pinnata
piperi Nelson	P. douglasii
planitiarum Nelson	P. andicola
pottsii Taylor, nomen nudum	P. paniculata
procumbens Lehmann	P. stolonifera × P. subulata
procumbens Sessé & Mociño	renaming of P. reticulata S. & M.
prostrata Aiton	P. stolonifera
puberula Nelson	P. longifolia
pulchella Loudon	P. paniculata
pulcherrimum Lundell	P. pilosa
pulchra Wherry	10–i
pyramidalis Smith	P. maculata
reflexa Sweet	P. maculata
reptans Michaux	P. stolonifera
reticulata Pavon	Loeselia coerulea
eticulata Sessé & Mociño	Loeselia coerulea
revoluta Aiken	P. glaberrima
richardsoni Hooker	10–iii
rigida Bentham	P. douglasii
rigida Shuttleworth ex Brand	P. floridana
riversii Taylor, nomen nudum	P. paniculata
roemeriana Scheele	10–ii
rosea Forbes	P. paniculata
rugelii Brand	P. amoena × P. divaricata
sabini Douglas ex Hooker	P. speciosa
scabra Sweet	P. paniculata
scleranthifolia Rydberg	P. diffusa
setacea Linnaeus	P. subulata
sevorsa Nelson	P. alyssifolia
shepherdii Young	P. paniculata
sibirica Linnaeus	10–iii
sickmanni Lehmann	P. paniculata
siebmanni Bentham	see P. sickmanni
speciosa Pursh	10–ii
speciosissima Maund	P. paniculata
spinosilla Sessé & Mociño	Loeselia mexicana
splendens Taylor, nomen nudum	P. paniculata
stansburyi (Torrey) Heller	10–i
stellaria Gray	P. bifida
stellaria Hort.	P. bifida × P. subulata
stolonifera Sims	10–i
suaveolens Aiton	P. maculata

...(continued)

Phlox

subulata Linnaeus	10–i
suffruticosa Hort.	P. maculata × P. carolina
suffruticosa Ventenat	P. carolina L.
suksdorfii St. John	P. speciosa
superba Brand	P. stansburyi
tardiflora Penny ex Young	P. maculata
tenera Pépin	P. paniculata
tenuifolia Nelson	10–ii
tenuis Nelson	P. cuspidata
texensis Lundell	P. nivalis
tharpii Whitehouse	P. glabriflora
thomsoni Courtois ex Steudel, nomen nudum	P. paniculata
tigrina Wenderoth ex Steudel, nomen nudum	P. paniculata
triflora Michaux	P. glaberrima
triovulata Thurber ex Torrey	10–ii
tumulosa Wherry	P. griseola
undulata Aiton	P. paniculata
unidentata Bertero ex Colla	Collomia biflora
uniflora Rafinesque	nomen dubium
vanhouttei Lindley	P. paniculata
variabilis Brand	10–i
verna Forbes	P. stolonifera
vernalis Salisbury	P. divaricata
vernoniana Loudon	P. paniculata
villosissima Whitehouse	P. pilosa
vincaefolia Pépin, nomen nudum	P. paniculata
violacea Sessé & Mociño	nomen dubium, perhaps Loeselia
virginica Loddiges ex Sweet, nomen nudum	P. paniculata
virginalis Carrière, nomen nudum	P. paniculata
virida Nelson	see P. viridis
viridis Nelson	P. longifolia
viscida Nelson	10–i
visenda Nelson	P. grayi
walteri Chapman	P. amoena Sims
wheeleriana Sweet	P. paniculata
wherryi Heath	P. pilosa
whitedii Nelson	P. speciosa
wilcoxiana Bogusch	P. drummondii
woodhousei Torrey ex Gray	P. speciosa
youngii Knight ex Paxton	P. paniculata

Phu

graeca Dodonaeus	Polemonium caeruleum
peregrinum Gesner	Polemonium caeruleum

Polemoniella

antarctica (Grisebach) Nelson & Macbride	Polemonium micranthum
antarcticum (Grisebach) Nelson & Macbride	see P. antarctica
gayana (Weddell) Nelson & Macbride	Polemonium micranthum

...*(continued)*

Polemoniella

gayanum (Weddell) Nelson & Macbride	see P. gayana
micrantha (Bentham) Heller	Polemonium micranthum

Polemonium

acaule Schiede ex Martens & Galeotti	Phacelia platycarpa (Hydrophyllaceae)
achilleaefolium Roemer & Schultes	Phacelia platycarpa (Hydrophyllaceae)
acutiflorum Willdenow ex Roemer & Schultes	P. caeruleum
albiflorum Eastwood	P. foliosissimum
album Fischer	P. caeruleum
amoenum Piper	P. carneum
antarcticum Grisebach	P. micranthum
archibaldae Nelson	P. foliosissimum
arcticum Nyl. & Saell.	P boreale
berryi Eastwood	P. pulcherrimum
bicolor Greenman	P. elegans
biflorum Kuntze	Collomia biflora
boreale Adams	6–i
brandegei Greene	6–ii
bursifolium Willdenow ex Roemer & Schultes	Phacelia platycarpa (Hydrophyllaceae)
caeruleum Linnaeus	6–i
·alifornicum Eastwood	6–i
calycinum Eastwood	P. californicum
caucasicum Busch	near P. boreale
campanulatum Fries	P. caeruleum
campanuloides Thunberg	Retzia campanuloides (Loganiaceae)
candidum Sessé & Mociño	Ipomopsis sp.
capitatum Bentham	P. boreale
capitatum Eschscholtz	Gilia capitata
carneum Gray	6–i
chartaceum Mason	6–ii
chinense Brand	P. caeruleum
ciliatum Willdenow ex Roemer & Schultes	Phacelia purshii Buckley (Hydrophyllaceae)
coeruleum, of Willdenow and many subsequent authors	see P. caeruleum
columbianum Rydberg	P. californicum
confertum Gray	6–ii
crassifolium (Bentham) Kuntze	Gilia crassifolia
dasyphylla, cited by Davidson as Polemonium	Phlox dasyphylla
decurrens Brand	P. foliosissimum
delicatum Rydberg	6–i
diminutum Klokov	P. caeruleum
dissectum Reichenbach	P. caeruleum
drummondii (Hooker) Kuntze	Phlox drummondii
dubium Linnaeus	Phacelia dubia (Hydrophyllaceae)
ehrenbergii Brand	P. grandiflorum
elatum Salisbury	P. caeruleum
elegans Greene	6–ii
eximium Greene	6–ii

. . . (continued)

Polemonium

fasciculatum Eastwood	P. pulcherrimum
filicinum Greene	P. foliosissimum
flavum Greene	6–i
foliolatum Klokov	P. caeruleum
foliosissimum Gray	6–i
frolovianum Fischer ex Herder	P. boreale
froloyianum Davidson	see P. frolovianum
gayanum (Weddell) Brand	P. micranthum
glabrum Davidson	6–i
gracile Douglas ex Lindley, nomen nudum	nomen dubium, Polemonium sp. from American arctic
gracile Fischer	P. caeruleum
grande Greene	P. foliosissimum
grandiflorum Bentham	6–i
grayanum Rydberg	P. viscosum
haydenii Nelson	P. pulcherrimum
helleri Brand	P. caeruleum
himalaicum H. W. W.	see P. himalayanum
himalayanum Baker	P. caeruleum
hinckleyi Standley	P. pauciflorum
hultenii Hara	P. humile Willdenow
humile Lindley	P. pulcherrimum
humile Salisbury	P. reptans
humile Willdenow	6–i
hyperborium Tolmatchew	P. boreale
incarnatum Gray ex Greene	P. carneum
incisum (Bentham) Kuntze	Gilia incisa
intermedium (Brand) Rydberg	P. caeruleum
jacobae Bergmans	garden hybrid of uncertain parentage
kiushianum Kitamura	P. caeruleum?
laciniatum (Ruiz & Pavon) Kuntze	Gilia laciniata
lacteum Lehmann	P. caeruleum
lanatum Pallas	P. boreale ·
lapponicum Klokov	near P. humile and P. boreale
lapponum Gandoger	P. boreale
laxiflorum Kitamura	P. caeruleum?
lemmonii Brand	P. viscosum
lindleyi Wherry	P. pulcherrimum
liniflorum Vassilev	near P. caeruleum
longii Fernald	P. reptans
luteum Greene	P. grandiflorum
luteum (Gray) Howell	P. carneum
macranthum (Chamisso) Klokov	near P. boreale
majus Tolmatchew	P. boreale?
mellitum Nelson	P. brandegei
mexicanum Cervantes ex Lagasca	6–i
mexicanum Nuttall	P. pulcherrimum
micranthum Bentham	6–iii
molle Greene	P. foliosissimum

...(continued)

Polemonium

montrosensis Nelson	P. pulcherrimum
morenonis Kuntze	Microsteris gracilis
moschatum Wormskjöld ex Graham	P. pulchellum
muricatum Lagasca ex Schrank	nomen dubium, not Polemonium
nevadense Wherry	P. pulcherrimum
nipponicum Kitamura	P. caeruleum?
nudipedum Klokov	near P. humile and P. pulchellum
nyctelea Linnaeus	Ellisia nyctelea (Hydrophyllaceae)
obscurum Blanco	Lepistemon flavescens (Convolvulaceae)
occidentale Greene	P. caeruleum
onegense Klokov	near P. pulchellum?
orbiculare Gandoger	P. pulcherrimum
oreades Gandoger	P. pulcherrimum
oregonense Gandoger	P. pulcherrimum
pacificum Vassilev	near P. caeruleum and P. boreale
paddoense Gandoger	P. californicum
parviflorum Tolmatchew	near P. caeruleum
parvifolium Nuttall ex Rydberg	P. pulcherrimum
pauciflorum Watson	6-i
pectinatum Greene	6-i
pilosum (Greenman) Jones	P. pulcherrimum
pimpinelloides Willdenow ex Roemer & Schultes	Phacelia platycarpa (Hydrophyllaceae)
prostratum Rudolph ex Georgi	P. boreale
pseudopulchellum Vassilev	near P. pulchellum
pterospermum (Bentham) Brand	P. caeruleum
pterospermum Nelson & Cockerell	P. foliosissimum
pulchellum Blytt	P. boreale
pulchellum Bunge	6-i
pulcherrimum Hooker	6-i
pumilum Gray ex Rydberg	see P. humile
quadriflorum Rafinesque	P. reptans
racemosum Kitamura	P. caeruleum?
reptans Linnaeus	6-i
rhaeticum Thomas	P. caeruleum
richardsonii Graham	P. boreale
robustum Rydberg	P. foliosissimum
roëlloides Thunberg	Retzia roëlloides (Loganiaceae)
rotatum Eastwood	P. pulcherrimum
rubrum Linnaeus	Ipomopsis rubra
samejedorum Gandoger	P. boreale
schizanthum Klokov	P. caeruleum
schmidtii Klokov	P. caeruleum
scopulinum Greene ex Rydberg	P. delicatum
shastense Eastwood	P. pulcherrimum
sibiricum Don	P. caeruleum
speciosum Fischer ex Hooker	P. boreale
speciosum Rydberg	P. viscosum
stenocalyx Standley	P. pauciflorum

... *(continued)*

Polemonium

tevesii Eastwood	P. californicum
tricolor Eastwood	P. californicum
vajgaczense Klokov	P. boreale × P. caeruleum
vanbruntiae Britton	P. caeruleum
valerianaefolia Gilibert	P. caeruleum
victoris Klokov	P. boreale × P. caeruleum
villosum Rudolph ex Georgi	P. caeruleum
villosum Sweet	P. boreale
viscosum Gray	P. elegans
viscosum Nuttall	6–ii
vulgare Gray	P. caeruleum
yezoënse (Miyabe & Kudo) Kitamura	P. caeruleum

Quamoclit

miscellaneous species	Quamoclit (Convolvulaceae)
pennatum Dillenius	Ipomopsis rubra

Rosenbergia

aschersoniana (Brand) House	Cobaea aschersoniana
campanulata (Hemsley) House	Cobaea campanulata
gracilis Oersted	Cobaea gracilis
macrostema (Pavon) House	Cobaea lutea
macrostoma (Pavon) House	see R. macrostema
minor (Martens & Galeotti) House	Cobaea minor
penduliflora Karsten	Cobaea penduliflora
pringlei House	Cobaea pringlei
scandens (Cavanilles) House	Cobaea scandens
stipularis (Bentham) House	Cobaea stipularis
trianaei (Hemsley) House	Cobaea trianaei
triflora (Smith) House	Cobaea triflora

Royenia

foliis acute dentatis Houstoun	Loeselia ciliata

Siphonella

miscellaneous species	Siphonella Small (Valerianella, Valeriana- ceae)
floribunda (Gray) Jepson	Linanthus nuttallii
montana Nuttall ex Gray	Linanthus nuttallii
nuttallii Heller	Linanthus nuttallii
parviflora Nuttall ex Gray	Linanthus nuttallii

Thouinia

miscellaneous species	Thounia L. (Linociera, Oleaceae), Thouinia Poit. (Sapindaceae), Thouinia Sm. (Humbertia, Convolvulaceae)
multifida Dombey ex Jussieu	Gilia laciniata

Tintinabulum

filiforme Rydberg	Gilia filiformis

Tunaria

albida Kuntze	Cantua pyrifolia

. . . (continued)

Valeriana

miscellaneous species	Valeriana L. (Valerianaceae)
coerulea Bauhin	Polemonium caeruleum
graeca Dodonaeus	Polemonium caeruleum
peregrina Dodonaeus	Polemonium caeruleum

Welwitschia

one species	Welwitschia Hook. (Gnetales)
densifolia (Bentham) Tidestrom	Eriastrum densifolium
diffusa (Gray) Rydberg	Eriastrum diffusum
filifolia (Nuttall) Rydberg	Eriastrum filifolium
floccosa (Gray) Rydberg	Eriastrum luteum
wilcoxii (Nelson) Rydberg	Eriastrum wilcoxii

CHAPTER VI

CHROMOSOME NUMBERS

The main outlines of the cytotaxonomy of the Polemoniaceae were established by Flory in 1937. In this paper Flory showed that the cytological characteristics of the tropical and temperate genera are conspicuously different; that the basic chromosome numbers in the temperate genera are $X = 7$, 8 and 9; and that each basic number tends to be constant within a major taxonomic group such as a genus or subgenus. Subsequent cytological explorations in the family have not only confirmed the principal conclusions reached by Flory but have in addition provided a firmer factual basis for them.

Two developments in particular have strengthened the cytotaxonomic analysis of the Phlox family. The first of these is a great increase in chromosome number records which are based, not on horticultural materials, as was the case in the pioneering studies, but on wild populations. The second is the construction of a more phylogenetic system of classification of the family. Consequently we now possess an enlarged body of data which can be organized in a more significant way than was possible previously.

The purposes of the present chapter are threefold: to record 148 new chromosome counts; to summarize all the known chromosome numbers in the family as based on a total of 380 counts; and to consider the meaning of these data in terms of trends in karyotype evolution within the family.

MATERIALS AND METHODS

Most of the chromosome counts made in our laboratory were obtained from squash preparations of pollen mother cells. The best results are secured by fixing the buds in 3 : 1 alcohol-propionic acid for several hours and subsequently transferring the fixed material to 70% alcohol. The PMC's are stained with propiono-carmine and mounted in Hoyer's medium according to the schedule given by Beeks (1955). The preparations can be studied most effectively by phase contrast. Drawings were made by camera lucida, the individual chromosomes or bivalents frequently being spread apart in the process of drawing.

The counts listed below are placed in the currently accepted nomenclature for each genus as presented in recent revisional or monographic studies. In a number of instances it has been necessary to make nomenclatural translations. Thus the *Phlox stellaria* Gray counted by Flory and Meyer is now treated as *Phlox bifida stellaria* (Gray) Wherry; and the *Polemonium molle* Greene reported on by Flory is now regarded as synonymous with *Polemonium foliosissimum* Gray.

The identification of the material counted presents an acute problem in a family containing as many taxonomically critical groups as the Polemoniaceae. The acuteness of this problem varies according to the particular group concerned. When an author reports 52 chromosomes in *Cobaea scandens*, a well known, widely cultivated and genetically uniform species, we can safely assume that the identification is correct. But in large areas of the family the identifications even of specialists do not have permanent value due to changes in the concepts concerning individual taxa. By retracing their steps it has been possible to infer that when Langlet and Sugiura counted $N = 9$ in *"Gilia laciniata"* they were dealing with *Gilia achilleaefolia*, and that the *"Gilia millefoliata"* counted by Langlet and Flory was actually *Gilia clivorum*. The report of tetraploidy in *"Phlox hoodii"* in 1937 could refer to any one of a dozen western American taxa or their hybrids.

The only real solution to this problem of identification is the preservation of a herbarium specimen of the plants which have been investigated cytologically. The identity of the material can then be verified by subsequent students in spite of changes in the nomenclature and the taxonomy. Counts which can be tied to a definite natural population also have more value botanically than counts obtained from garden plants of ambiguous origin and frequently mixed parentage. The geographical source of the plants is cited in the following list wherever known. A special effort was made through correspondence with several earlier authors of chromosome counts to bring out this information in cases where it was not published in the original report. However, in many cases, particularly in the genus *Phlox*, the true identity of the material counted is now a matter of guesswork, and some of the early cytological work will consequently have to be done over again.

In the list which follows, newly reported counts are denoted by an asterisk (*). Chromosome number reports published prior to 1930 are ignored. A taxonomic species containing both diploid and tetraploid forms, or tetraploid plus higher polyploid forms, is counted as more than one species for purposes of computing percentages of polyploidy. Voucher specimens, where available, are mostly filed in the herbaria of the Rancho Santa Ana Botanic Garden and the University of California at Berkeley; a specimen of A. Nygren is in the Institute of Plant Systematics and Genetics, Uppsala. The sequence of data in the list is as follows:

Taxon; (synonym); **somatic number.**

Source of material; author; *herbarium specimen* (if any); figure (if any).

LIST OF CHROMOSOME NUMBERS

1. Cobaea

1. **C. scandens** Cav.; **2N = 52.**
    Commercial nursery, native in Mexico; Flory 1937; Janaki-Ammal in Darlington and Wylie 1955; *Grant; *Grant 18258.*

2. Cantua

1. **C. buxifolia** Lam.; **2N = c. 54.**
    Commercial nursery, native in Peru; Flory 1937.
2. ***C. candelilla** Brand; **2N = 54.**
    Tarata, Andes, Peru; Grant; *Grant 2275* (original collection by H. Johnson); Fig. 62b.
3. ***C. pyrifolia** Lam.; **2N = 54.**
    Cultivated in Vavra Estate, Los Angeles, California, native in Peru; Grant; *Grant 2245.*

3. Loeselia

1. ***L. mexicana** (Lam.) Brand; **2N = 18.**
    Santa María del Rio, San Luis Potosí, Mexico; Grant; *J. Rzedowski 3213*; Fig. 62a.

4. Polemonium

I. Section Polemonium

1. **P. boreale** Adams; (listed thus and as *P. richardsonii* and *P. lanatum humile*, synonyms); **2N = 18.**
    Botanic gardens, native in circumboreal region; Flory 1937, Griesinger 1937.
    Isfjorden, Spitzbergen; Flovik 1940.
2a. **P. caeruleum** L. ssp. **caeruleum**; (listed as *P. caeruleum*, *P. caeruleum gracile*, *P. caeruleum himalayanum*, and *P. sibiricum* Don., synonyms); **2N = 18.**

Botanic gardens, native in Eurasia; Clausen 1931, Flory 1937, Griesinger 1937.

2b. **P. caeruleum** ssp. **vanbruntiae** (Britt.) Davidson; (listed as *P. vanbruntiae*, a synonym); 2N = 18, 36.
Probably West Virginia; 2N = 18; Flory 1937.
Beaverkill, west of Kingston, Catskill Mts., Ulster Co., New York; 2N = 36; Berger in Cain 1944: 462; *Wm. Bonisteel in 1940.*

2c. **P. caeruleum** ssp. **villosum** (Rud.) Brand; (listed as *P. acutiflorum* Willd., a synonym); 2N = 18.
Maskjok, District of Tana, Finnmark, Norway; Nygren in Löve and Löve 1948: 86; *A. Nygren in 1947.*

3. **P. californicum** Eastw.; 2N = 36.
Silver Lake, Sierra Nevada, Amador Co., California; Davidson 1950; *L. Constance 3086.*

4. **P. carneum** Gray; 2N = 18.
Botanic gardens, native in Pacific North America; Clausen 1931, Griesinger 1937.
Stonybrook Canyon, Alameda Co., California; Davidson 1950.

5. **P. foliosissimum** Gray; (listed thus and as *P. filicinum* Greene and *P. molle* Greene, synonyms); 2N = 18.
Botanic gardens, native in western United States; Clausen 1931, Flory 1937, Griesinger 1937, Davidson 1950.
*Hobble Creek, Wasatch Mts., Utah Co., Utah; Grant; *Grant 9694*; Fig. 62c.

6. **P. mexicanum** Cerv. ex Lag.; 2N = 18.
Botanic gardens, native in central Mexico; Clausen 1931, Flory 1937.

7. **P. pauciflorum** Wats.; 2N = 18.
Botanic gardens, native in northern Mexico; Clausen 1931, Griesinger 1937, *Grant and Latimer; *Grant 3039.*

8. **P. pulcherrimum** Hook.; 2N = 18.
Botanic garden, native in western North America; Clausen 1931.
Rossland, British Columbia; Davidson 1950.
Eagle Rock, Modoc Co., California; Davidson 1950.

9. **P. reptans** L.; 2N = 18.
Botanic gardens, native in eastern North America; Clausen 1931, Griesinger 1937.
Bruceton Mills, east of Morgantown, Appalachian Mts., Preston Co., West Virginia; Flory 1937.

II. Section Melliosma

10. **P. confertum** Gray; **2N = 18.**
    Source unspecified, native in Rocky Mt. region; Flory 1937.
11. **P. viscosum** Nutt.; **2N = 18.**
    Botanic garden, native in Rocky Mt. region; Griesinger 1937.
    *Medicine Bow Mts., Albany Co., Wyoming; Grant; *Grant 9726.*

III. Section Polemoniastrum

12. ***P. micranthum** Benth.; **2N = 18.**
    Rio Mayo, Gob. Comodoro Rivadavia, Argentina; Grant; *Grant 2724* (original seed collection by A. Soriano).

5. Allophyllum

1a.***A. divaricatum** (Nutt.) Grant, Sierra Nevada Race; **2N = 16.**
    Groveland, Tuolumne Co., California; Grant; *D. D. Keck 6456, Grant 9450.*
    Twain Harte, Tuolumne Co., California; Grant and Beeks; *Grant 9449*; Fig. 62e.
    Hetch Hetchy Reservoir, Tuolumne Co., California; Grant and Beeks; *Grant 2205.*
1b.***A. divaricatum,** Coast Range Race; **2N = 16.**
    Tassajara Springs, Monterey Co., California; Grant; *Grant 9146.*
    China Camp, Santa Lucia Mts., Monterey Co., California; Grant; *Grant 9144.*
1c.***A. divaricatum,** San Gabriel Race; **2N = 18.**
    Claremont, Los Angeles Co., California; Grant; *Grant 9150.*
    Cow Canyon, San Gabriel Mts., Los Angeles Co., California; Grant; *Grant 9400, 16008.*
2a.***A. integrifolium** (Brand) Grant, High Montane Race; **2N = 16.**
    Lake Alpine, Sierra Nevada, Calaveras Co., California; Grant and Beeks; *Grant 9436.*
2b.***A. integrifolium,** Mid-Altitude Race; **2N = 18.**
    Kern River, Sierra Nevada, Tulare Co., California; Grant and Beeks; *Grant 9263.*
3a.***A. glutinosum** (Benth.) Grant, San Diego Race; **2N = 18.**
    Santa Ana Mts., Orange Co., California; Grant; *Grant 8659.*
3b.***A. glutinosum,** Mountain Top Race; **2N = 18.**
    Santa Rosa Mt., Riverside Co., California; Grant; *Grant 9395*; Fig. 62d.
4. ***A. gilioides** (Benth.) Grant; **2N = 18.**

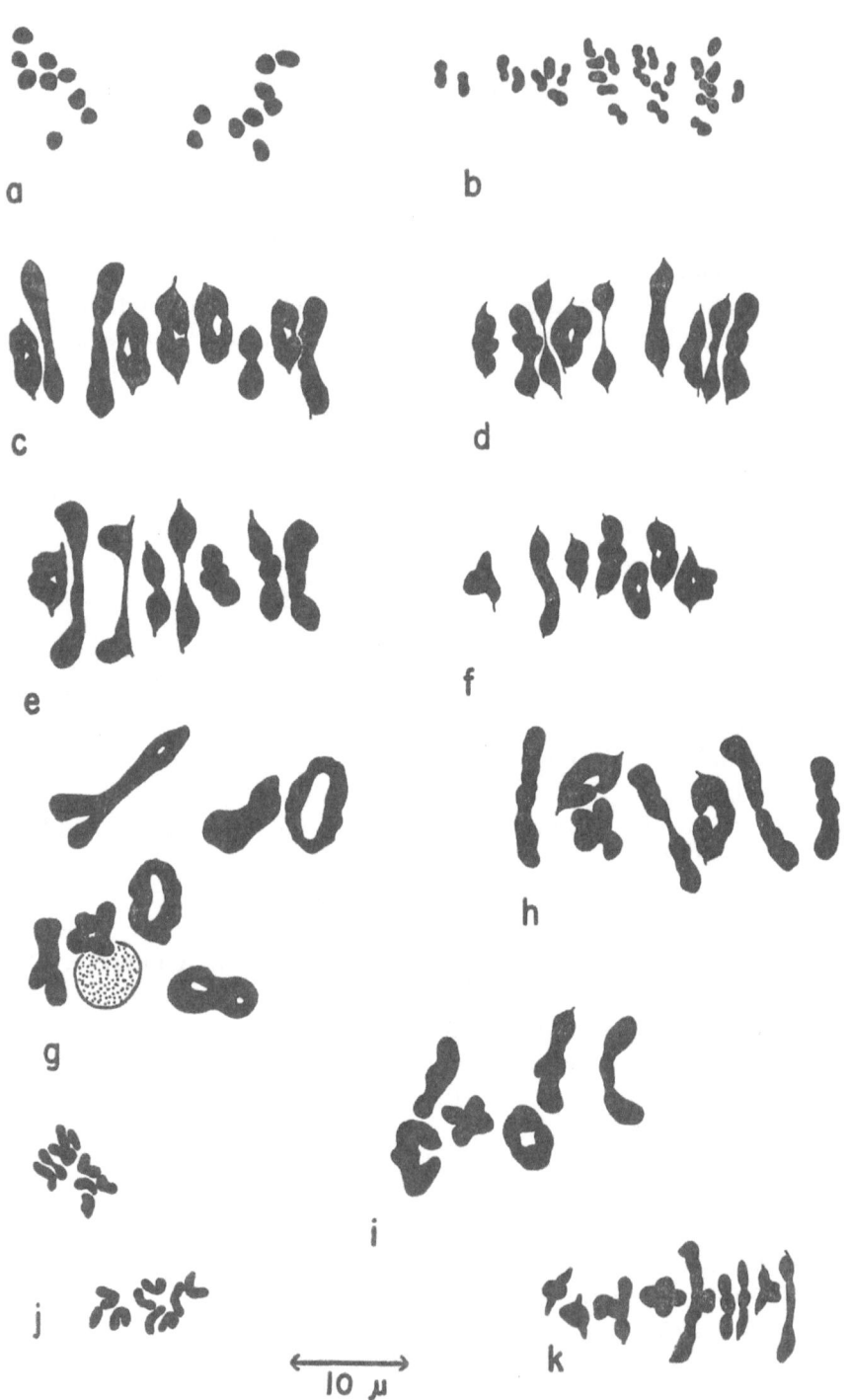

Fig. 62. Meiotic chromosomes of representative Polemoniaceae.
a) Loeselia mexicana; AI, 9–9. *b*) Cantua candelilla; 27 II. *c*) Polemonium foliosissimum; 9 II.
d) Allophyllum glutinosum; 9 II. *e*) Allophyllum divaricatum; 8 II. *f*) Ipomopsis congesta; 7 II.
g) Ipomopsis pumila; 7 II. *h*) Ipomopsis aggregata attenuata; 7 II. *i*) Gymnosteris parvula, 6 II.
j) Leptodactylon californicum; AI, 9–9. *k*) Navarretia atractyloides; 9 II. (original, from camera
ucida drawings of PMC's.)

Pinnacles National Monument, San Benito Co., California; Grant; *Grant 8712.*

5. **A. violaceum* (Heller) Grant; 2N = 18.
    Mt. Pinos, Ventura Co., California; Grant and Beeks; *Grant 9416.*

6. Collomia

1. **C. cavanillesii** Hook. & Arn.; (listed as *C. coccinea* Lehm., a synonym); 2N = 32.
    Source unspecified, native in Chile; Sugiura 1936, Flory 1937.
2. **C. grandiflora** Dougl.; 2N = 16.
    Source unspecified, native in western North America; Flory 1937.
3. **C. heterophylla** Hook; 2N = 16.
    Source unspecified, native in Pacific North America; Flory 1937.
    *Kings Mt., San Mateo Co., California; Grant; *Grant 8921.*
    *Mill Valley, Marin Co., California; Grant; *Grant 8879.*
4. **C. linearis** Nutt.; 2N = 16.
    Source unspecified, native in North America; Sugiura 1936, Flory 1937.

7. Gymnosteris

1. **G. nudicaulis** (Hook. & Arn.) Greene; (listed as *G. nudicaulis pulchella* (Greene) Brand, a synonym); 2N = 12.
    Steamboat Springs, Washoe Co., Nevada; McMillan 1949; *H. L. Mason 13983.*
2. **G. parvula* (Rydb.) Heller; 2N = 12.
    Wagontire, Harney Co., Oregon; Grant and Latimer; *Grant 9806;* Fig. 62i.

8. Phlox

I. Section Phlox

1. **P. adsurgens** Torr. ex Gray; 2N = 14, c. 21.
    Commercial nursery, native in northern California and Oregon; 2N = 14; Flory 1937.
    Horticultural form from a commercial nursery; 2N = c. 21, this triploid form may be a garden hybrid; Flory 1934.
2. **P. amplifolia** Britt.; 2N = 14.
    Source unspecified, native in eastern United States; Flory 1937, Meyer 1944.
3a. **P. bifida** Beck ssp. **bifida**; 2N = 14.

Commercial nursery, native in eastern United States; Flory 1937, Meyer 1944.

3*b*. **P. bifida** ssp. **stellaria** (Gray) Wherry; (listed as *P. stellaria* Gray, a synonym); **2N = 14.**
Commercial nursery, native in eastern United States; Flory 1934, Meyer 1944.

4. **P. buckleyi** Wherry; **2N = 28.**
Kate's Mt., White Sulphur Springs, Greenbrier Co., West Virginia; Flory 1934, Meyer 1944.
Lewisburg, Greenbrier Co., West Virginia; Meyer 1944.

5*a*. **P. carolina** L. ssp. **carolina**; (listed as *P. carolina heterophylla*, a synonym); **2N = 14 + 0—2 B.**
Source unspecified, native in southeastern United States; Meyer 1944.

5*b*. **P. carolina** ssp. **alta** Wherry; (listed as *P. suffruticosa* Vent., and *P. glaberrima suffruticosa*, synonyms); **2N = 14 + 0—1 B, and 21.**
Commercial nursery, native in southeastern United States; vars. Miss Verboom, Princess Ingrid, Rosalinda, Miss Lingard had 2N = 14; var. Rosalinda showed supernumeraries in certain individuals; in one instance Miss Lingard had 2N = 21; in attempting to understand the different chromosome numbers found by different observers in Miss Lingard it is important to bear in mind that this cultivar is not a simple member of any native taxon but is probably a hybrid; Flory 1934, Meyer 1944.

6. **P. glaberrima** L.; **2N = 14 + 1—3 B.**
Source unspecified, native in eastern United States; Meyer 1944.

7a. **P. maculata** L. ssp. **maculata**; **2N = 14.**
Commercial nursery, native in eastern North America; Flory 1934.

7*b*. **P. maculata** ssp. **pyramidalis** (Smith) Wherry; (listed as *P. longiflora* Hort., a synonym); **2N = 14.**
Commercial nursery, native in eastern North America; Flory 1937.

8. **P. ovata** L.; **2N = 14.**
Commercial nursery, native in eastern United States; Flory 1934.

9. **P. paniculata** L.; **2N = 14 + 0—10 B.**
Shenandoah River, Clarke Co., Virginia; 2N = 14; Flory 1934.
Commercial nursery, native in eastern United States; typical form and 126 horticultural varieties had 2N = 14; Flory 1934, 1937, Meyer 1944.
Commercial nursery; 40 varieties had 2N = 14 + 1—6 B; vars. Hodur, Mrs. Jenkins, Mrs. Jenner had 2N = 14 + 7—8 B; vars. Commander and Mme. Louise Abbema had 2N = 14 + 10 B; Meyer 1944.

10. **P. ✕ procumbens** Lehm.; $(= P.\ stolonifera \times subulata)$; $2N = 14 + 0—1\ B$.
    Commercial nursery; Flory 1934.
11. **P. pulchra** Wherry; (listed as *P. ovata*.var. *pulchra*, a synonym); $2N = 14$.
    Source unspecified, probably from the type locality, Oakman, Walker
    Co., Alabama; Meyer 1944.
12. **P. stolonifera** Sims; (listed thus and as *P. procumbens coerulea*, a syno-
    nym); $2N = 14$.
    Commercial nursery, native in eastern United States; Flory 1934,
    Meyer 1944.
13. **P. subulata** L.; $2N = 14 + 0—13\ B,\ and\ 28$.
    Woodstock, Shenandoah Co., Virginia; $2N = 14$; Meyer 1944.
    Commercial nursery; the typical form and 15 horticultural varieties
    had $2N = 14$; Flory 1934, Meyer 1944.
    Commercial nursery; var. *cuspidata* had $2N = 14 + 0—6\ B$; one
    plant of the typical form had $2N = 14 + 0—13\ B$; 4 plants of the
    typical form had $2N = 28$; Meyer 1944.

II. Section Divaricatae

14. **P. amoena** Sims; $2N = 14$.
    Chocolocco Mt., Calhoun Co., Alabama; Flory 1934.
    Commercial nursery, native in southeastern United States; Meyer 1944.
15. **P. ✕ arendsii** Hort.; $(= P.\ divaricata \times paniculata)$; $2N = 14$.
    Commercial nursery; Flory 1934, Meyer 1944.
16. **P. divaricata** L.; $2N = 14 + 0—1\ B$.
    Greenbrier Co., West Virginia; Flory 1934.
    Private garden in Winchester, Virginia; Flory 1934.
    Source unspecified, native in eastern United States; 9 individuals
    without and one individual with a supernumerary; Meyer 1944.
17. **P. drummondii** Hook.; $2N = 14$.
    Commercial nursery, native in Texas; several forms; Flory 1934,
    Meyer 1944.
18. **P. nivalis** Sweet; $2N = 14 + 0—4\ B$.
    Clarksville, Mecklenburg Co., Virginia; Flory 1937, Meyer 1944.
    Rockingham, Richmond Co., North Carolina; Flory 1937, Meyer 1944.
    Chapel Hill, Orange Co., North Carolina; Flory 1937, Meyer 1944.
    Oxford, Granville Co., North Carolina; Flory 1937, Meyer 1944.
    Wake Forest, Wake Co., North Carolina; Flory 1937, Meyer 1944.
    Rock Hill, York Co., South Carolina; Flory 1937, Meyer 1944.
    Monroe, Walton Co., Georgia; Flory 1937, Meyer 1944.
    Stone Mountain, Dekalb Co., Georgia; Flory 1937, Meyer 1944.

19. **P. pilosa** L.; (listed thus and also as *P. argillacea* Clute & Ferriss, a synonym); 2N = 14.
Chocolocco Mt., Calhoun Co., Alabama; Flory 1934.
Commercial nursery, native in eastern North America; Flory 1934, Meyer 1944.

III. Section Occidentales

20. **P. diffusa** Benth.; 2N = 28.
Commercial nursery, native in western North America; Flory 1934.
21. **P. douglasii** Hook.; 2N = 14.
Commercial nursery, native in northwestern North America; Flory 1934.
22. **P. hoodii** Hort.; (= *P. hoodii* Rich.?); 2N = 28.
Commercial nursery, native in western North America; Flory 1937.

9. Microsteris

1a. **M. gracilis** (Hook.) Greene ssp. **gracilis**; (listed thus and as *Phlox gracilis* (Hook.) Greene, a synonym); 2N = 14.
Source unspecified; Mason 1941.
*Coram, Flathead Co., Montana; Kinch 1956; *D. Dunn 10492*.
*Whitefish, Flathead Co., Montana; Kinch 1956; *D. Dunn 9382*.
*Estancia Quichaura, Tecka, Terr. Chubut, Argentina; 2N = c. 14; Kinch 1956; *Kinch* (original seed collection by A. Soriano).
1b.*M. gracilis** ssp. **humilis** (Greene) Grant; 2N = 14.
Logan, Cache Co., Utah; Kinch 1956; *Kinch* (original seed collection by R. Shaw).

10. Gilia

I. Section Giliastrum

1. *G. campanulata** Gray; 2N = 18.
Fish Lake Valley, Esmeralda Co., Nevada; Grant and Latimer; *Grant 9831*.
2. *G. filiformis** Parry; 2N = 18.
Bradbury Well, Death Valley, Inyo Co., California; Grant; *Grant 17535*.
3. **G. foetida** Gill. ex. Benth.; 2N = 18.
Uspallata, Depto. Las Heras, Prov. Mendoza, Argentina; Covas and Schnack 1946.
4. *G. incisa** Benth.; 2N = 18.

Blewett, Uvalde Co., Texas; Grant and Latimer; *B. Turner 3869, Grant 3007.*

Austin, Travis Co., Texas; Grant and Latimer; *Grant 3005* (original seed collection by B. Turner).

Between San Luis Potosí and Rioverde, San Luis Potosí, Mexico; Grant and Latimer; *J. Rzedowski 3626, Grant 3012.*

5. ***G. latifolia** Wats.; **2N = 36.**

Mecca, Riverside Co., California; Grant; *Grant 17519.*

6. **G. rigidula** Benth.; **2N = 18, 36.**

Source unspecified; 2N = 18; Flory 1937.

*Austin, Travis Co., Texas; 2N = 36; Grant and Latimer; *Grant 9570.*

*Devils Backbone, near San Marcos, Hayes Co., Texas; 2N = 36; Grant and Latimer; *Grant 9569.*

*Carta Valley, Edwards Co., Texas; 2N = 36; Grant and Latimer; *Grant 3009* (original seed collection by B. Turner).

*East of Langtry, Valverde Co., Texas; 2N = 36 ; Grant and Latimer; *Grant 3010* (original seed collection by B. Turner).

II. Section Giliandra

7. ***G. hutchinsifolia** Rydb.; **2N = 18.**

Fish Lake Valley, Esmeralda Co., Nevada; Grant and Latimer; *Grant 9826*; Fig. 63a.

8. ***G. micromeria** Gray; **2N = 16, 34, 50.**

Benton, Mono Co., California; 2N = 16; Grant and Latimer; *Grant 9618*; Fig. 63b.

Short Canyon, Kern-Inyo Cos., California; 2N = 34; Grant and Latimer; *Grant 9338;* Fig. 63c.

Red Rock Canyon, Kern Co., California; 2N = 34; Grant; *Grant 17575.*

Toyabe Mts., Nye Co., Nevada; 2N = 34; Grant and Latimer; *Grant 9825.*

Winnemucca, Humboldt Co., Nevada; 2N = 50; Grant and Latimer; *Grant 9821*; Fig. 63d.

9. ***G. pentstemonoides** Jones; **2N = 16.**

Blue Mesa, Cimarron, Montrose Co., Colorado; Grant and Latimer; *Wm. A. Weber in 1955, Grant 3019* (from Weber's collection).

10. ***G. pinnatifida** Nutt.; **2N = 16.**

Creede, Mineral Co., Colorado; Grant and Latimer; *Grant 9470.*

Fort Garland, Costilla Co., Colorado; Grant and Latimer; *Grant 9471.*

Roggen, Weld Co., Colorado; Grant and Latimer; *Grant 9504.*

11. ***G. subnuda** Gray; **2N = 16.**

  Mesa Verde, Montezuma Co., Colorado; Grant and Latimer; *Grant 9465.*

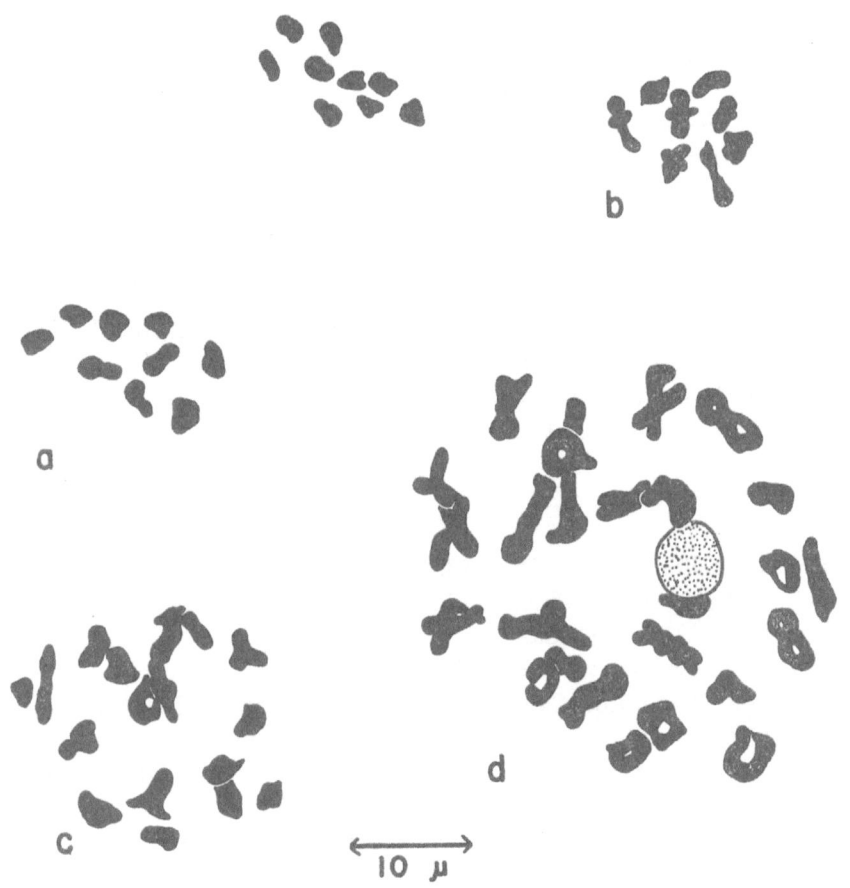

Fig. 63. Dibasic polyploidy in the Gilia leptomeria group.
a) Gilia hutchinsifolia; N = 9. *b*) diploid G. micromeria; N = 8. *c*) tetraploid G. micromeria; N = 17. *d*) hexaploid G. micromeria; N = 25. (original, from camera lucida drawings of PMC's; scale same as Fig. 62.)

III. Section Gilia

12a. **G. achilleaefolia** Benth. ssp. **achilleaefolia**; (listed thus and as *G. laciniata* Ruiz & Pavon due to misidentification); **2N = 18.**

  San Luis Obispo, San Luis Obispo Co., California; Grant 1954a; *Grant 8557.*

Moraga Canyon, Contra Costa Co., California; Grant 1954a; *Grant 8872.*

Commercial nursery, identification confirmed by the present author; Flory 1937.

Botanic gardens; Langlet 1936, Sugiura 1940. (These authors reported their counts under the name *Gilia laciniata.* I have obtained seeds of what is passing as *Gilia laciniata* in several European botanic gardens. When grown to maturity this material turns out to be typical *G. achilleaefolia.* Wild populations of *G. achilleaefolia* are known to be 9-paired, whereas no wild population of *G. laciniata* with N = 9 has yet been found. Herbarium specimens of the strains on which Langlet and Sugiura are presumed to have based their counts are preserved as *Grant 2288* and *2289.*)

12b. **G. achilleaefolia** ssp. **multicaulis** (Benth.) Grant; (listed as *Gilia achilleaefolia,* a synonym); 2N = 18.

Tres Pinos, San Benito Co., California; Grant 1954a; *Grant 8433.*
Kings Mt., San Mateo Co., California; Grant 1954a; *Grant 8903.*
Corte Madera Ridge, Mt. Tamalpais, Marin Co., California; Grant 1954a; *Grant 8878.*

13. **G. angelensis** Grant; 2N = 18.

Lytle Canyon, San Gabriel Mts., San Bernardino Co., California; Grant 1952b; *Grant 8661.*
Wheeler Hot Springs, Ventura Co., California; Grant 1952b; *Grant 8706.*

14a. **G. capitata** Sims ssp. **capitata**; 2N = 18.

Commercial nursery; Flory 1934.
Mayacama Mts., Napa Co., California; Grant 1950; *Grant 7702, 7985.*
Hatchet Mt., Shasta Co., California; Grant 1950; *Grant 8023.*

14b. **G. capitata** ssp. **abrotanifolia** (Nutt.) Grant; 2N = 18.

Kaweah River, Sierra Nevada, Tulare Co., California; Grant 1952a; *Grant 8481B.*
Wheeler Hot Springs, Ventura Co., California; Grant 1952a; *Grant 8705.*
Mentone, San Bernardino Mts., San Bernardino Co., California; Grant 1952a; *Grant 8699.*

14c. **G. capitata** ssp. **chamissonis** (Greene) Grant; 2N = 18.

Point Reyes Peninsula, Marin Co., California; Grant 1950; *Grant 7662.*

14d. **G. capitata** ssp. **mediomontana** Grant; 2N = 18.

Yosemite Valley, Mariposa Co., California; Grant 1950; *Grant 8610.*

14e. **G. capitata** ssp. **pacifica** Grant; 2N = 18.

Cape Mendocino, Humboldt Co., California; Grant 1950; *J. P. Tracy 18225.*

14f. **G. capitata** ssp. **pedemontana** Grant; **2N = 18.**

Calaveras River, Sierra Nevada, Calaveras Co., California; Grant 1950; *Grant 8498.*

Kaweah River, Sierra Nevada, Tulare Co., California; Grant 1950; *Grant 8481A.*

14g. **G. capitata** ssp. **staminea** (Greene) Grant; **2N = 18.**

Antioch, Contra Costa Co., California; Grant 1950; *H. L. Mason 12566, Grant 7925.*

14h. **G. capitata** ssp. **tomentosa** (Eastw.) Grant; **2N = 18.**

Tomales Bay, Marin Co., California; Grant 1950; *Grant 7991.*

15. **G. clivorum** (Jeps.) Grant; (listed thus and as *G. millefoliata* Fisch. & Mey. formerly a synonym); **2N = 36.**

Grizzly Peak, near Berkeley, Alameda Co., California; Grant 1954b; *Grant 8536.*

Strawberry Canyon, near Berkeley, Alameda Co., California; Grant 1954b; *Grant 8535.*

Moraga Canyon, Contra Costa Co., California; Grant 1954b; *Grant 8647.*

Panoche Pass, San Benito Co., California; Grant 1954b; *Grant 8435.*

*Temecula, Riverside Co., California; Grant and Beeks; *Grant 9315.*

Botanic gardens, as *Gilia millefoliata*, with which *G. clivorum* was formerly confused; Langlet 1936, Flory 1937.

16. **G. laciniata** Ruiz & Pavon; **2N = 36.**

Cerro San Gerónimo, Lima, Peru; Grant 1954c; *Grant 2244* (original seed collection by M. Reiche and O. Velarde).

17. **G. millefoliata** Fisch. & Mey.; **2N = 18.**

Point Reyes Peninsula, Marin Co., California; Grant 1954b; *Grant 8419.*

18. ***G. nevinii** Gray; **2N = 36.**

Whites Landing, Santa Catalina Island, California; Grant and Latimer; *Grant 9209.*

19a. **G. tricolor** Benth. ssp. **tricolor**; **2N = 18.**

Botanic gardens; Langlet 1936, Sugiura 1936, Flory 1937.

Byron, Contra Costa Co., California; Grant 1952b; *Grant 8640.*

19b. **G. tricolor** ssp. **diffusa** (Congdon) Mason & Grant; **2N = 18.**

Parkfield, Fresno Co., California; Grant 1952b; *D. D. Keck 6307.*

Tipton, San Joaquin Valley, Tulare Co., California; Grant 1952b; *Grant 8654.*

Kern Canyon, Sierra Nevada, Kern Co., California; Grant 1952b; *Grant 8466.*

20. ***G. valdiviensis** Griseb.; 2N = 18.
Quebrada de las Palmas de Alvarado, Limache, Prov. Valparaiso, Chile; Grant; *Grant 2988* (original seed collection by A. Garaventa).
Rio Claro, Fundo El Colorado, Itahue, Prov. Talca, Chile; Grant; *Grant 2989* (original seed collection by A. Garaventa).

IV. Section Arachnion

21. **G. aliquanta** Grant ssp. **aliquanta**; 2N = 18.
Red Rock Canyon, Kern Co., California; Grant, Beeks and Latimer 1956; *Grant 9117*.
Pear Blossom, Mojave Desert, Los Angeles Co., California; Grant, Beeks and Latimer 1956; *Grant 9324*.
Desert Springs, Mojave Desert, San Bernardino-Los Angeles Cos., California; Grant, Beeks and Latimer 1956; *D. D. Keck 6261*.

22a. **G. brecciarum** Jones ssp. **brecciarum**; 2N = 18.
Halleluja Junction, Plumas Co., California; Grant, Beeks and Latimer *Grant 9669*.
Ballinger Canyon, Cuyama Valley, Santa Barbara Co., California; Grant, Beeks and Latimer 1956; *Grant 2682*.
Mt. Pinos, Ventura Co., California; Grant, Beeks and Latimer 1956; *Grant 16049, 9409*.
*Mt. Pinos; Grant and Latimer; *Grant 9773A, 9773B*.
*Walker River north of Sonora Junction, Mono Co., California; Grant and Latimer; *Grant 9672*.
*Topaz Lake, Douglas Co., Nevada; Grant and Latimer; *Grant 9780, 9781, 9782*.
*South 'of Carson City, Douglas-Ormsby Cos., Nevada; Grant and Latimer; *Grant 9787*.
*North of Carson City, Ormsby Co., Nevada; Grant and Latimer; *Grant 9789*.
*Austin, Lander Co., Nevada; Grant and Latimer; *Grant 9823*.

22b. **G. brecciarum** ssp. **argusana** Grant; 2N = 18.
Searles Station, El Paso Mts., Kern Co., California; Grant, Beeks and Latimer 1956; *Grant 8856*.
Randsburg, El Paso Mts., Kern Co., California; Grant, Beeks and Latimer 1956; *Grant 9358*.
Jawbone Canyon near Red Rock Canyon, Kern Co., California; Grant, Beeks and Latimer 1956; *E. K. Balls 8569*.
Homewood Canyon, Argus Mts., Inyo Co., California; Grant, Beeks and Latimer 1956; *Grant 9347*.

22c. **G. brecciarum** ssp. **neglecta** Grant; **2N = 18.**

Short Canyon, Kern-Inyo Cos., California; Grant, Beeks and Latimer 1956; *Grant 9335*.

23a. **G. cana** (Jones) Heller ssp. **cana; 2N = 18.**

Rock Creek, Sierra Nevada, Mono Co., California; Grant, Beeks and Latimer 1956; *Grant 9249*.

23b. **G. cana** ssp. **bernardina** Grant; **2N = 18.**

Cactus Flat, San Bernardino Mts., San Bernardino Co., California; Grant, Beeks and Latimer 1956; *Grant 8686*.

*Lucerne Valley, San Bernardino Co., California; Alva Grant; *Grant 9948B*.

23c. **G. cana** ssp. **speciosa** (Jeps.) Grant; **2N = 18.**

Short Canyon, Kern-Inyo Cos., California; Grant, Beeks and Latimer 1956; *Grant 8860*.

23d. **G. cana** ssp. **triceps** (Brand) Grant; **2N = 18.**

China Lake, Mojave Desert, San Bernardino Co., California; Grant, Beeks and Latimer 1956; *Grant 9341*.

Homewood Canyon, Argus Mts., Inyo Co., California; Grant, Beeks and Latimer 1956; *Grant 9344*.

Trona, Mojave Desert, Inyo Co., California; Grant, Beeks and Latimer 1956; Grant 9342, 9343.

Wildrose Canyon, Panamint Mts., Inyo Co., California; Grant, Beeks and Latimer 1956; *Grant 8821, 9351*.

Bradbury Well, Death Valley, Inyo Co., California; Grant, Beeks and Latimer 1956; *Grant 9072*.

24. **G. crassifolia** Benth.; **2N = 36, 72.**

Puente del Inca, Depto. Las Heras, Prov. Mendoza, Argentina; 2N = 36; Grant, Beeks and Latimer 1956; *J. Hunziker 6316*.

Estancia Quichaura, Tecka, Terr. Chubut, Argentina; 2N = 72; Grant, Beeks and Latimer 1956; *Grant 2294, 2959* (original seed collection by A. Soriano).

*Cañadon Leon, Terr. Santa Cruz, Argentina; 2N = 72; Grant and Latimer; *A. Soriano 5081, Grant 2978* (from Soriano seed collection).

25. **G. diegensis** (Munz) Grant; **2N = 18.**

Cuyamaca Mts., San Diego Co., California; Grant, Beeks and Latimer 1956; *Grant 9225, 9227*.

Anza Junction, Riverside Co., California; Grant, Beeks and Latimer 1956; *Grant 9234, 9235, 9236*.

Temecula, Riverside Co., California; Grant, Beeks and Latimer 1956; *Grant 9316*.

*Palomar Mt., San Diego Co., California; Grant and Latimer; *Grant 2953, 2954.*

26.   *G. inconspicua** (Smith) Sweet; (previously reported counts under this name apply to *G. interior*); 2N = 36.

Lizard Butte, near Marsing, Canyon-Owyhee Cos., Idaho; Grant and Latimer; *Grant 9814.*

West of Marsing, Owyhee Co., Idaho; Grant and Latimer; *Grant 9816.*

Blue Mt. Pass, Malheur Co., Oregon; Grant and Latimer; *Grant 9819.*

Sparks, Washoe Co., Nevada; Grant and Latimer; *Grant 9795.*

South of Carson City, Douglas Co., Nevada; Grant and Latimer; *Grant 9785.*

Virginia City, Ormsby-Storey Cos., Nevada; Grant; *H. L. Mason 13988.*

Kyle Canyon, Charleston Mts., Clark Co., Nevada; Grant and Latimer; *Grant 9952.*

Beaverdam Mts., Washington Co., Utah; Grant and Latimer; *Grant 9968, 9972.*

Kramer Hills, San Bernardino Co., California; Grant and Latimer; *Grant 9902.*

Mesquite, Mohave Co., Arizona; N = c. 18; Grant and Latimer; *Grant 9963.*

27.   **G. interior** (Mason & Grant) Grant; (listed as *G. inconspicua*, a synonym); 2N = 18.

Kern Canyon, Sierra Nevada, Kern Co., California; Grant, Beeks and Latimer 1956; *H. L. Mason 14004.*

Hobo Springs, Kern Canyon, Sierra Nevada, Kern Co., California; Grant, Beeks and Latimer 1956; *Grant 9108.*

Cuyama Valley, Santa Barbara Co., California; Grant, Beeks and Latimer 1956; *Grant 9100, 9361, 9362.*

Ballinger Canyon, Cuyama Valley, Santa Barbara Co., California; Grant, Beeks and Latimer 1956; *Grant 2674, 2678, 2680, 9365.*

Walker River near Sonora Junction, Mono Co., California; Grant, Beeks and Latimer 1956; *Grant 9673.*

*Topaz Lake, Douglas Co., Nevada; Grant and Latimer; *Grant 9779.*

*Carson City, Ormsby Co., Nevada; Grant and Latimer; *Grant 9788.*

28a.   **G. latiflora** Gray ssp. **latiflora**; 2N = 18.

Adelanto, Mojave Desert, San Bernardino Co., California; Grant, Beeks·and Latimer 1956; *Grant 8663.*

Apple Valley, Mojave Desert, San Bernardino Co., California; Grant, Beeks and Latimer 1956; *Rancho Santa Ana Botanic Garden 20683.*

28b.   **G. latiflora** ssp. **cuyamensis** Grant; 2N = 18.

Lockwood Valley, Ventura Co., California; Grant, Beeks and Latimer 1956; *Grant 9420.*

28c. ***G. latiflora** ssp. **davyi** (Miillken) Grant; **2N = 18.**

Gorman, Los Angeles Co., California; Grant and Latimer; *Grant 9578.*

28d. **G. latiflora** ssp. **excellens** (Brand) Grant; **2N = 18.**

Johannesburg, El Paso Mts., Kern Co., California; Grant, Beeks and Latimer 1956; *Grant 9221.*

29a. **G. leptantha** Parish ssp. **leptantha; 2N = 18.**

Upper Santa Ana River, San Bernardino Mts., San Bernardino Co., California; Grant, Beeks and Latimer 1956; *Grant 9155.*

29b. **G. leptantha** ssp. **pinetorum** Grant; **2N = 18.**

Mt. Pinos, Ventura Co., California; Grant, Beeks and Latimer 1956; *Grant 16047, 16052.*

29c. **G. leptantha** ssp. **purpusii** (Milliken) Grant; **2N = 18.**

Kern River near Johnsondale, Sierra Nevada, Tulare Co., California; Grant, Beeks and Latimer 1956; *Grant 9264.*

Kern River near Old Isabella, Sierra Nevada, Kern Co., California; Grant, Beeks and Latimer 1956; *Grant 9220.*

Greenhorn Mts., Sierra Nevada, Kern Co., California; Grant, Beeks and Latimer 1956; *Grant 9218.*

29d. **G. leptantha** ssp. **transversa** Grant; **2N = 18.**

Cajon Pass, San Bernardino Mts., San Bernardino Co., California; Grant, Beeks and Latimer 1956; *Grant 9385.*

*Lucerne Valley, San Bernardino Co., California; Grant and Latimer; *Grant 9945.*

29e. **G. leptantha** ssp. **vivida** Grant; **2N = 18.**

Big Pines, San Gabriel Mts., Los Angeles Co., California; Grant, Beeks and Latimer 1956; *Grant 16055.*

30. **G. minor** Grant; **2N = 18.**

Kramer Junction, Mojave Desert, San Bernardino Co., California; Grant, Beeks and Latimer 1956; *Grant 8851, 9222.*

Homewood Canyon, Argus Mts., Inyo Co., California; Grant, Beeks and Latimer 1956; *Grant 9350.*

Gorman, Los Angeles Co., California; Grant, Beeks and Latimer 1956; *Grant 9092.*

Ballinger Canyon, Cuyama Valley, Santa Barbara Co., California; Grant, Beeks and Latimer 1956; *Grant 2677, 2684, 9098.*

Cuyama Valley, Santa Barbara Co., California; Grant, Beeks and Latimer 1956; *Grant 9099.*

Wickenburg, Maricopa Co., Arizona; Grant, Beeks and Latimer 1956; *Grant 9298.*

31. **G. modocensis** Eastw.; (listed as *G. sinuata*, previously considered synonymous, and as *G. tetrabreccia*, previously considered distinct); 2N = 36.

> Cajon Pass, San Bernardino Mts., San Bernardino Co., California; Grant, Beeks and Latimer 1956; *Grant 15993.*
>
> Phelan, Mojave Desert, San Bernardino Co., California; Grant, Beeks and Latimer 1956; *Grant 2951, 15995.*
>
> Highway 138 near Big Pines Junction, Los Angeles-San Bernardino Cos., California; Grant, Beeks and Latimer 1956; *D. D. Keck 6262, Grant 9318.*
>
> Kern River south of Johnsondale, Sierra Nevada, Tulare Co., California; Grant, Beeks and Latimer 1956; *Grant 9265.*
>
> Walker River north of Sonora Junction, Mono Co., California; Grant, Beeks and Latimer 1956; *Grant 9672A.*
>
> Halleluja Junction, Plumas Co., California; Grant, Beeks and Latimer 1956; *Grant 9668.*
>
> Mt. Pinos, Ventura Co., California; Grant, Beeks and Latimer 1956; *Grant 16042.*
>
> *Mt. Pinos; Grant and Latimer; *Grant 9770, 9771.*
>
> *Doyle, Lassen Co., California; Grant and Latimer; *Grant 9666A.*
>
> *Jess Valley, Warner Mts., Modoc Co., California; Grant and Latimer; *Grant 9804.*
>
> *Topaz Lake, Douglas Co., Nevada; Grant and Latimer; *Grant 9784.*
>
> *Austin, Lander Co., Nevada; Grant and Latimer; *Grant 9822.*
>
> *Big Pines, San Gabriel Mts., Los Angeles Co., California; Grant and Latimer; *Grant 9845, 9846, 9848.*

32a. **G. ochroleuca** Jones ssp. **ochroleuca**; 2N = 18.

> Short Canyon, Kern-Inyo Cos., California; Grant, Beeks and Latimer 1956; *Grant 8858.*

32b. **G. ochroleuca** ssp. **bizonata** Grant; 2N = 18.

> Stauffer Junction near Mt. Pinos, Ventura Co., California; Grant, Beeks and Latimer 1956; *Grant 16040.*

32c. **G. ochroleuca** ssp. **exilis** (Gray) Grant; 2N = 18.

> Dripping Springs, Riverside Co., California; Grant, Beeks and Latimer 1956; *Grant 9317.*
>
> Fulmore Lake, Riverside Co., California; Grant, Beeks and Latimer 1956; *Grant 16107.*
>
> Whitewater Canyon near San Gorgonio Pass, Riverside Co., California; Grant, Beeks and Latimer 1956; *Grant 9031.*

33a. **G. ophthalmoides** Brand ssp. **ophthalmoides**; (listed thus and as *G. o. clokeyi*, with which previously confused) ; 2N = 36.

Westgard Pass, White Mts., Inyo Co., California; Grant, Beeks and Latimer 1956; *Grant 9431.*

Kyle Canyon, Charleston Mts., Clark Co., Nevada; Grant Beeks and Latimer 1956; *E. K. Balls and R. Straw 19276.*

*Deep Springs, Inyo Co., California; Grant and Latimer; *Grant 9839.*

*Sweetwater Mts., Mono Co., California; Grant and Latimer; *P. A. Munz 21118, Grant 2948* (from Munz seed collection).

*Beaverdam Mts,. Washington Co., Utah; Grant and Latimer; *Grant 9969.*

*Dinosaur National Monument, Uintah Co., Utah; Grant and Latimer; *Grant 9705.*

33b. **G. ophthalmoides** ssp. **australis** Grant; **2n = 36.**

Canyon del Oro, west of Tucson, Pima Co., Arizona; Grant, Beeks and Latimer 1956; *K. F. Parker 7428, Grant 2728* (from Parker seed collection).

33c. **G. ophthalmoides** ssp. **flavocincta** (Nelson) Grant; **2N = 36.**

Apache Trail, Maricopa-Gila Co., Arizona; Grant, Beeks and Latimer 1956; *Grant 9307.*

Wickenburg, Maricopa Co., Arizona; Grant, Beeks and Latimer 1956; *Grant 9297.*

34. **G. sinuata** Dougl.; **2N = 36.**

Box S Springs, north base of San Bernardino Mts., San Bernardino Co., California; Grant, Beeks and Latimer 1956; *Grant 8691.*

Adelanto, Mojave Desert, San Bernardino Co., California; Grant, Beeks and Latimer 1956; *Grant 8662.*

El Paso Mts., Kern Co., California; Grant, Beeks and Latimer 1956; *Grant 9557.*

Benton Station, Mono Co., California; Grant, Beeks and Latimer 1956; *Grant 9613.*

Benton, Mono Co., California; Grant, Beeks and Latimer 1956; *Grant 9620.*

Doyle, Plumas Co., California; Grant, Beeks and Latimer 1956; *Grant 9666.*

*Morongo Valley, Riverside Co., California; Grant and Latimer; *Grant 9895.*

*Lucerne Valley, San Bernardino Co., California; Grant and Latimer; *Grant 9946.*

*Rock Creek, Mono Co., California; Grant and Latimer; *Grant 9776.*

*Carson City, Ormsby Co., Nevada; Grant and Latimer; *Grant 9663.*

*Columbia River at junction of highways 395 and 730, Umatilla Co., Oregon; Grant and Latimer; *Grant 9811.*

*Shoshone, Lincoln Co., Idaho; Grant and Latimer; *Grant 9742.*

35. **G. tenuiflora** Benth, ssp. **tenuiflora; 2N = 18.**

Arroyo Seco, Monterey Co., California; Grant, Beeks and Latimer 1956; *G. L. Stebbins 3945.*

36. **G. transmontana** (Mason & Grant) Grant; **2N = 36.**

Short Canyon, Kern-Inyo Cos., California; Grant, Beeks and Latimer 1956; *Grant 9337.*

Johannesburg, San Bernardino-Kern Cos., California; Grant, Beeks and Latimer 1956; *Grant 8847.*

Mountain Pass, east of Baker, Mojave Desert, San Bernardino Co., California; Grant, Beeks and Latimer 1956; *Grant 9060.*

Cuyama Valley, Santa Barbara Co., California; Grant, Beeks and Latimer 1956; *Grant 8696.*

*Kramer Junction, San Bernardino Co., California; Grant and Latimer; *Grant 9906.*

*Deep Springs, Inyo Co., California; N = c. 18; Grant and Latimer; *Grant 9838.*

37. **G. tweedyi** Rydb.; **2N = 36.**

Encampment, Carbon Co., Wyoming; Grant, Beeks and Latimer 1956; *Grant 9725.*

*Shirley Basin, Carbon Co., Wyoming; Grant and Latimer; *Grant 9727.*

V. Section Saltugilia

38. **G. australis** (Mason & Grant) Grant; **2N = 18.**

Morongo Canyon, Riverside Co., California; Grant and Grant 1954; *Grant 15985.*

39. **G. capillaris** Kell.; **2N = 18.**

Sonora Pass, Sierra Nevada, Tuolumne Co., California; Grant and Grant 1954; *Grant 9185.*

40. **G. caruifolia** Abrams; **2N = 18.**

Palomar Mt., San Diego Co., California; Grant and Grant 1954; *Grant 16018.*

41a. **G. leptalea** (Gray) Greene ssp. **leptalea; 2N = 18.**

Lake Almanor, Plumas Co., California; Grant and Grant 1954; *Grant 8925.*

*Ebbetts Pass, Sierra Nevada, Alpine Co., California; Grant and Beeks; *Grant 9438.*

41b.*G. **leptalea** ssp. **bicolor** Mason & Grant; **2N = 18.**

Lake Alpine, Calaveras-Alpine Cos., California; Grant and Beeks; *Grant 9437.*

Carson Pass, Sierra Nevada, Alpine Co., California; Grant and Beeks; *Grant 9434.*

42. *G. scopulorum Jones; 2N = 18, 36.

Sheephole Mts., Mojave Desert, San Bernardino Co., California; 2N = 18; Grant; *Grant 17525.*

Amargosa Mts., Mojave Desert, Inyo Co., California; 2N = 36; Grant; *Grant 17531.*

43a. G. splendens Dougl. ssp. splendens; 2N = 18.

Morongo Canyon, Riverside Co., California; Grant and Grant 1954 *Grant 15984.*

43b. G. splendens ssp. grantii (Brand) Grant; 2N = 18.

Cow Canyon, San Gabriel Mts., Los Angeles Co., California; Grant and Grant 1954; *Grant 16006.*

44. *G. stellata Heller; 2N = 18.

Morongo Canyon, Riverside Co., California; Grant; *Grant 15986.*

11. Ipomopsis

I. Section Phloganthea

1. *I. multiflora (Nutt.) Grant; 2N = 14, 28.

Mt. Lemmon, Pima Co., Arizona; 2N = 14; Grant and Latimer; *Grant 16127.*

Prescott, Yavapai Co., Arizona; 2N = 28; Grant and Latimer; Grant 9453.

2. *I. polyantha (Rydb.) Grant; 2N = 14.

Pagosa Springs, Archuleta Co., Colorado; Grant and Latimer; *Grant 9468.*

Rowes Well, Grand Canyon, Coconino Co., Arizona; Grant and Latimer; *Grant 9459.*

3. *I. tenuifolia (Gray) Grant; 2N = 14.

Jacumba, San Diego Co., California; Grant; *E. K. Balls 18903.*

II. Section Ipomopsis

4a. I. aggregata (Pursh) Grant ssp. aggregata; (also listed as *Gilia aggregata* Benth., a synonym); 2N = 14.

Source unspecified, native in western North America; Flory 1937.

*Metolius River, west of Bend, Deschutes Co., Oregon; Grant; *P. Grun 37.*

*Red Rock, near Diamond Mt., Plumas Co., California; Grant.

*Yellow Jacket Pass, Rocky Mts., Archuleta Co., Colorado; Grant and Latimer; *Grant 9467.*

*Williams, Coconino Co., Arizona; Grant and Latimer; *P. A. Munz 18245.*

4b.*I. aggregata ssp. arizonica (Greene) Grant; 2N = 14.

Charleston Mts., Clark Co., Nevada; Grant and Latimer; *Grant 9520.*

Grand View Point, Grand Canyon, Coconino Co., Arizona; Grant and Latimer; *Grant 9460.*

Zion Canyon, Kane Co., Utah; Grant and Latimer; *Grant 9517.*

4c.*I. aggregata ssp. attenuata (Gray) Grant; 2N = 14.

Rabbit Ears Pass, Rocky Mts., Routt-Jackson Co., Colorado; Grant and Latimer; *Grant 9501.*

Sonora Pass, Sierra Nevada, Tuolumne-Mono Co., California; Grant and Latimer; *Grant 9539*; Fig. 62h.

4d.*I. aggregata ssp. bridgesii (Gray) Grant; 2N = 14.

Ilillouette Creek, Yosemite National Park, Mariposa Co., California; Grant and Latimer; *Grant 9530.*

Peppermint Creek, above Camp Nelson, Sierra Nevada, Tulare Co., California; Grant and Latimer; *Grant 9525.*

4e.*I. aggregata ssp. candida (Rydb.) Grant; 2N = 14.

Boulder Canyon, Rocky Mts., Boulder Co., Colorado; Grant and Latimer *Grant 9491.*

Turkey Creek Canyon, southwest of Morrison, Rocky Mts., Jefferson Co., Colorado; Grant and Latimer; *Grant 9478.*

5. I. longiflora (Torr.) Grant; (also listed as *Gilia longiflora* (Torr) Don, a synonym); 2N = 14.

Source unspecified, native in southwestern United States; Flory 1937.

*Rio Grande Valley, Socorro Co., New Mexico; Grant; *Grant 8813* (original seed collection by O. Norwell).

*Roggen, Weld Co., Colorado; Grant and Latimer; *Grant 9503.*

6. I. rubra (L.) Wherry; (listed as *Gilia rubra* (L.) Heller, a synonym) 2N = 14.

Source unspecified, native in southeastern United States; Flory 1937.

7. *I. tenuituba (Rydb.) Grant; 2N = 14.

Paragonah, Iron Co., Utah; Grant and Latimer; *Grant 9685.*

8. *I. thurberi (Torr.) Grant; 2N = 14.

Santa Rita Mts., Pima Co., Arizona; Grant; *Grant 16132.*

III. Section Microgilia

9a.*I. congesta (Hook). Grant ssp. congesta; 2N = 14.

Bridgeport, Mono Co., California; Grant and Latimer; *Grant 9674.*

West of Vernal, Uintah Co., Utah; Grant and Latimer; *Grant 9701*.

West of Maybell, Moffat Co., Colorado; Grant and Latimer; *Grant 9709*.

Big Lost River, Bingham Co., Idaho; Grant and Latimer; *Grant 9737*; Fig. 62f.

9b. *I. congesta ssp. montana (Nels. & Kennedy) Grant; 2N = 14.

Cedar Mt., Warner Range, Modoc Co., California; Grant; *Grant 8944*.

10. *I. frutescens (Rydb.) Grant; 2N = 14.

Zion Canyon, Kane Co., Utah; Grant and Latimer; *Grant 9518*.

11. *I. gossypifera (Gill.) Grant; 2N = 14.

Las Casitas, Dep. Iglesia, Prov. San Juan, Argentina; Hunziker and Caso; *J. Hunziker and O. Caso 4743*; Fig. 64.

12. *I. minutiflora (Benth.) Grant; 2N = 14.

Snake River, Baker-Malheur Cos., Oregon; Grant and Latimer; *Grant 9813*.

13. *I. polycladon (Torr.) Grant; 2N = 14.

Valley Wells, Mojave Desert, San Bernardino Co., California; Grant and Beeks; *P. A. Munz 17413*.

Fish Lake Valley, Esmeralda Co., Nevada; Grant and Latimer; *Grant 9827*.

Fig. 64. Somatic chromosomes of Ipomopsis gossypifera. (original, by J. Hunziker and O. Caso; × 3200.)

14. *I. pumila (Nutt.) Grant; 2N = 14.

Duchesne, Duchesne Co., Utah; Grant and Latimer; *Grant 3029, 9697*.

Big Horn Basin, Big Horn Co., Wyoming; Grant and Latimer; *Grant 3028* (original seed collection by A. Beetle); Fig. 62g.

15. *I. roseata (Rydb.) Grant; 2N =28.

Colorado National Monument, Mesa Co., Colorado; Grant and Latimer; *Grant 9511*.

16a. *I. spicata (Nutt.) Grant ssp. spicata; 2N = 14.

Boulder, Boulder Co., Colorado; Grant and Latimer; *Grant 9505*.

16b. *I. spicata ssp. capitata (Gray) Grant; 2N = 14.

Hoosier Ridge, Park Range, Park-Summit Cos., Colorado; Grant and Latimer; *Grant 9486*.

12. Eriastrum

1. *E. densifolium (Benth.) Mason ssp. austromontanum (Craig) Mason; 2N = 14.

San Jacinto Mts., Riverside Co., California; Grant; *Grant 16088*.

2. *E. sapphirinum (Eastw.) Mason ssp. dasyanthum (Brand) Mason;
2N = 14.
Claremont, Los Angeles Co., California; Grant; *Grant 16001*.
3. E. virgatum (Benth.) Mason; (listed as *Hugelia virgata* Benth., a synonym); 2N = 14.
Source unspecified, native in California; Flory 1937.

13. Langloisia

1. *L. schottii (Torr.) Greene; 2N = 14.
Morongo Valley, Riverside Co., California; Grant; *Grant 15988*.
Deep Springs, Inyo Co., California; Grant and Latimer; *Grant 9840*.
2. *L. setosissima (Torr. & Gray) Greene; 2N = 14.
Deep Springs, Inyo Co., California; Grant and Latimer; *Grant 9835*.

14. Navarretia

1. *N. atractyloides (Benth.) Hook. & Arn.; 2N = 18.
Claremont, Los Angeles Co., California; Grant and Latimer; *Grant
9609*; Fig. 62k.
2. *N. hamata Greene; 2N = 18.
Cobal Canyon, San Gabriel Mts., Los Angeles Co., California; Grant
and Latimer; *Grant 9844*.
3. *N. minima Nutt.; 2N = 18.
Source unrecorded, native in Pacific North America; Alva Grant.
4. *N. pauciflora Mason; 2N = c. 18.
Lower Lake, Lake Co., California; Alva Grant.

15. Leptodactylon

1. L. californicum Hook. & Arn.; 2N = 18.
Source unspecified, native in California; Flory 1937.
*Santa Ana Canyon, Orange Co., California; Grant; *Rancho Santa
Ana Botanic Garden 4453*; Fig. 62j.

16. Linanthus

1. L. androsaceus (Benth.) Greene ssp. micranthus (Steudel) Mason;
(listed as *L. parviflorus* Greene, a synonym); 2N = 18.
Source unspecified, native in California; Flory 1937.
2. L. aureus (Nutt.) Greene; 2N = 18.
Source unspecified, native in southwestern North America; Flory 1937.
3. L. dianthiflorus (Benth.) Greene; 2N = 18.
Source unspecified, native in California; Flory 1937.

4. **L. dichotomus** Benth.; 2N = 18.
    Source unspecified, native in southwestern North America; Flory 1937.
5. **L. grandiflorus** (Benth.) Greene; (listed as *L. densiflorus* Milliken
    a synonym); 2N = 18.
    Source unspecified, native in California; Flory 1937.

KARYOTYPE EVOLUTION

One hundred and thirty-two species of Polemoniaceae, or 41% of the species in the family, have now been determined for chromosome number. Counts are available for each one of the five tribes and for all but two of the genera. The Phlox family is thus fairly well known cytologically, better known in fact than the majority of angiospermous families or groups of comparable size. The information now on hand is adequate to establish certain conclusions concerning the general pattern of karyotype evolution in this family.

Nevertheless, some conspicuous gaps remain in our body of data. It may be worthwhile, before proceeding with the discussion, to indicate the places where further cytotaxonomic work is needed. Whereas the minor genera *Allophyllum* and *Microsteris* and the larger genera *Gilia* and *Ipomopsis* have been rather thoroughly investigated, the tropical genera *Huthia* and *Bonplandia* are still completely unknown cytologically, and *Cobaea* and *Loeselia* are only known from single counts. Counts are available for a fairly large proportion of the species of *Polemonium*, *Collomia* and *Phlox*, but these counts are unevenly distributed among the different sections of the genera, and can be related to specific wild populations in relatively few instances. No chromosome counts at all are available for the most primitive section, *Collomiastrum*, of *Collomia*, and only three counts for the large and complex section *Occidentales* of *Phlox*. Much more work remains to be done in *Eriastrum*, *Navarretia* and *Linanthus*.

The adequacy of the existing information on chromosome numbers depends on the uses to which this information is put. The available data are reliable for discussions of basic chromosome number but only roughly indicative for estimates of the incidence of polyploidy. This is because we are dealing with taxa in which little aneuploidy but a fair amount of irregularly distributed euploidy exists. A small sample is sufficient to establish the basic number of such a taxon, whereas a much more complete sampling is necessary to reveal accurately the proportion of polyploid species.

The basic numbers of the genera of Polemoniaceae are summarized in Table 10. Where the haploid numbers of a group are evidently polyploid and the original basic number can only be inferred, Tischler (1954) has designated the former as "B" and the latter as "X". Following this practice, the B numbers of *Cobaea* and *Cantua* are 26 and 27 respectively, and their common denomi-

Table 10. Basic Chromosome Numbers of the Genera of Polemoniaceae

Tribe Cobaeeae
   1. Cobaea   B = 26; X = 9?

Tribe Cantueae
   2. Cantua   B = 27; X = 9.
   3. Huthia   ?

Tribe Bonplandieae
   4. Bonplandia ?
   5. Loeselia   X = 9.

Tribe Polemonieae
   6. Polemonium X = 9.
   7. Allophyllum X = 9, 8.
   8. Collomia   X = 8.
   9. Gymnosteris X = 6.
  10. Phlox   X = 7.
  11. Microsteris   X = 7.

Tribe Gilieae
  12. Gilia   X = 9, 8.
  13. Ipomopsis   X = 7.
  14. Eriastrum   X = 7.
  15. Langloisia   X = 7.
  16. Navarretia   X = 9.
  17. Leptodactylon X = 9.
  18. Linanthus   X = 9.

nator is X = 9. The basic number of X = 9 is found also in *Loeselia*, in the most primitive genus in the Polemonium tribe (*Polemonium*), and in the most primitive genus in the Gilia tribe (*Gilia*). In each of the five tribes, therefore, the basic chromosome number is 9, and this number may be regarded as basic also for the family.

The basic numbers of X = 8, 7 and 6 which are associated in certain genera with advanced morphological and ecological characteristics are almost certainly derived from the original basic number of 9 by reduction. This reduction series is most clearly displayed in the Polemonium tribe. As shown in Chapter 7, the most primitive genus in this tribe is *Polemonium* with X = 9. One derivative genus, *Allophyllum*, includes both 9-and 8-paired species, and another derivative genus, *Collomia*, consists so far as we now know only of 8-paired species and their tetraploids. *Gymnosteris* with X = 6 is a highly specialized group allied to *Collomia*. *Phlox* (X = 7) is clearly allied to, but more specialized than, *Collomia* (X = 8), and *Microsteris* with X = 7 is an offshoot from *Phlox*.

As pointed out by Grant (1956a), "This reduction in chromosome number is in general correlated with reduction in morphology, simplification of the life cycle, and increase in uniformity in the breeding system." Thus *Microsteris* with N = 7 consists of ephemeral self-pollinating uniform annuals, and *Gymnosteris* with N = 6 possesses at once the lowest chromosome number

and the most highly reduced life cycle known in the family, the plants being
diminutive leafless self-pollinating annuals with a homogeneous, even static,
pattern of variation. These endpoints in the reduction series are related
through intermediate states, as represented by *Phlox* (N = 7) and *Collomia*
(N = 8), to the 9-paired genus *Polemonium* with its long-lived perennial
habit, cross-fertilizing breeding system, and high level of variability.

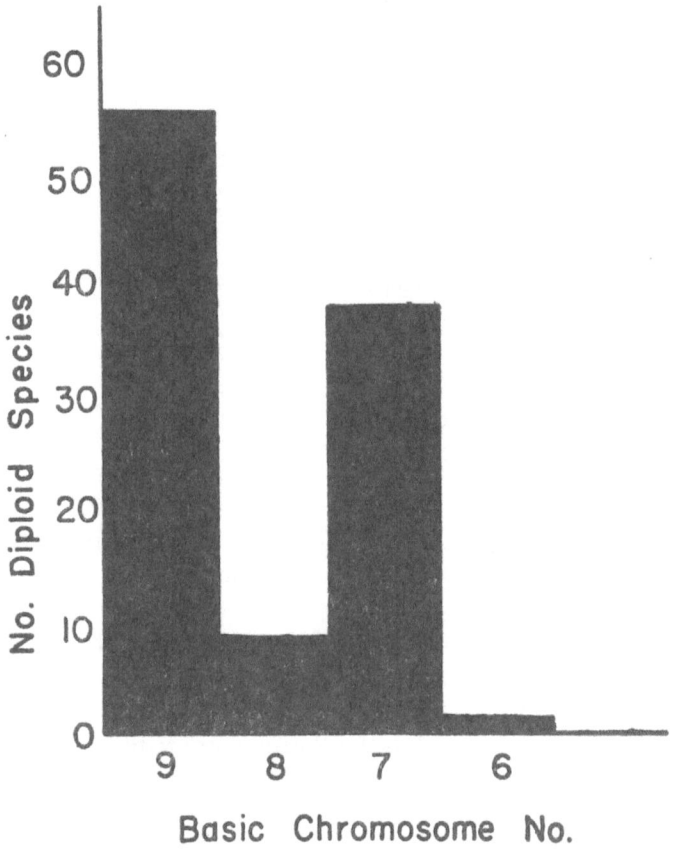

Fig. 65. Frequency distribution of basic chromosome numbers in the Phlox family.

In the Gilia tribe these correlations are not so evident. There are sections
of *Ipomopsis* and of *Gilia* whose general morphological characters have remained at approximately the same primitive level of advancement. In the
former X = 7 and in the latter X = 9. On the other hand, we find X = 9
in *Linanthus*, which has attained a degree of advancement comparable to that
of *Phlox* (X = 7) in the Polemonium tribe. It is probable that the basic

numbers of the different phyletic lines in the Gilia tribe were established early in their history and have remained constant during the course of many subsequent evolutionary changes. Nevertheless, it is interesting that the reduction from $X = 9$ to $X = 7$ has occurred independently in both temperate zone tribes.

The frequency distribution of the various basic numbers presents an interesting picture. As shown in Figure 65 two peaks exist in the family, one corresponding to $N = 9$ and the other to $N = 7$, while the intermediate class of $N = 8$ is represented by a minority of species. That this distribution is real and not an artifact of our sampling is indicated by a comparison between the direct tally incorporated into Figure 65 and the estimated total number of diploid species in the entire family with a given haploid number. The latter figures consist of the probable number of diploid species in each genus or section, grouped according to the known basic number of that taxon. The probable number of diploid species is estimated in turn from the known proportion of polyploids in each genus or section. The percentage frequency of diploid species with different haploid numbers as estimated by the indirect method agrees very closely – within one or two percent – with the frequency distribution of percentages obtained from the actual tally.

The basic number of $X = 8$ is poorly represented in each of the two temperate zone tribes. In the Polemonium tribe it is found only in *Collomia*, which is a relatively small genus, and in the minor genus *Allophyllum*. Until recently no 8-paired species was known at all in the Gilia tribe. This number then turned up in a few species of the section *Giliandra* of *Gilia*.

The causes of this curious distribution of chromosome numbers are unexplainable at the present time. For the angiosperms as a whole $N = 8$ is one of the most common chromosome numbers. In the Polemoniaceae, however, it seems to represent an adaptive valley bounded by the peaks of 9 and 7.

Aneuploid differences in the Polemoniaceae are found chiefly in the higher taxonomic categories. Whole genera are uniform in respect to the basic chromosome number, whereas related genera frequently differ in this regard. This situation is propitious for taxonomy at the genus level but is decidedly inconvenient for cytogenetics. There are, however, two areas known in the family where aneuploidy occurs within a comparium or group of intercrossable species.

Both 9- and 8-paired species, and indeed 9- and 8-paired races of the same species, exist in *Allophyllum* (Fig. 62d, e). These species and races can be intercrossed with ease and their hybrids are semifertile. Unpublished data show that the hybrids derived from heteroploid crosses are about as fertile as those derived from homoploid crosses. Unfortunately, owing to the extraordinary difficulty of obtaining good cytological preparations of pollen mother cells in *Allophyllum*, it has not been possible to analyze these hybrids cyto-genetically. The other species group containing both 9- and 8-paired species is the *Gilia leptomeria-hutchinsifolia* complex (Fig. 63). The existence of natural occurring 17- and 25-paired polyploids in the same complex indicates that the diploid species will form hybrids. These species are highly specialized desert annuals, however, which are not easily grown in the experimental garden. It may be a long time before the mechanism of chromosome number reduction in the Polemoniaceae can be elucidated.

In comparison with many angiospermous families there is relatively little polyploidy in the Polemoniaceae. The 31 polyploid species so far discovered comprise 23% of the species counted. This average percentage for the family as a whole embraces a wide variety of conditions ranging from *Allophyllum* with no known polyploidy to *Cantua* with no known diploid.

The proportion of polyploid species in the individual genera is as follows: *Cobaea* and *Cantua*, 100%; *Gilia*, 32%; *Phlox*, 27%; *Collomia*, 25%; *Polemonium*, 15%; *Ipomopsis*, 12%; and *Loeselia*, *Allophyllum*, *Gymnosteris*, *Microsteris*, *Eriastrum*, *Langloisia*, *Navarretia*, *Leptodactylon*, and *Linanthus*, 0%.

As noted in an earlier paragraph these estimates are less reliable in a poorly sampled genus, like *Linanthus*, than in a thoroughly sampled one such as *Allophyllum*. As a matter of fact, it would be surprising if some polyploidy is not revealed by future studies in *Eriastrum*, *Navarretia* and *Linanthus*. Nevertheless, when all allowances are made for the imperfections of our present estimates, we are still faced with real differences between certain genera in the frequency of polyploidy.

Similar differences exist between the sections of a single genus. In *Phlox* for example, very little polyploidy has been found in the herbaceous perennial groups, but the western montane cespitose phloxes may well prove to be a polyploid complex. In *Gilia* the greatest amount of polyploidy has been found in the desert section *Arachnion* (44%) and the least in the woodland section *Saltugilia* (12%).

By far the most commonly encountered level of ploidy in the family is tetraploidy. Triploidy has been reported in two instances in Phlox, but inasmuch as the plants involved were obtained from nursery sources the most likely assumption is that they were spontaneous garden hybrids of diploids and tetraploids. Hexaploidy is found in *Cantua*, in one species of *Gilia* (*G. micromeria*, and in a modified form in *Cobaea*. The highest level of ploidy so far discovered is octoploidy in *Gilia crassifolia* from Argentina.

The distribution of the higher polyploids throughout the family is not what one might expect from the assumption that an ample supply of genomes is conducive to the building up of higher ploidy levels. In the North American deserts, for example, there are six genomically well differentiated tetraploid species in the section *Arachnion* of *Gilia*. They occur sympatrically with one another and with several related diploid species in various combinations. A concerted search for hexaploid and octoploid forms in this group, either in nature or as spontaneous derivatives of artificial hybrids in the experimental garden, has so far proven fruitless. Yet an octoploid is present in Argentina where the same section is represented by only one or two tetraploid species, and, conversely, a hexaploid Gilia belonging to the small section *Giliandra* is found in the North American deserts. Evidently the presence of a diversity of genomes is not a major limiting factor in the development of polyploid types in this group.

A voluminous discussion has appeared in the literature during recent years concerning the factors which favor the appearance and perpetuation of polyploidy in higher plants. Correlations have been traced between this phenomenon and latitude, elevation, climate, stage of succession in the habitat, nutritional conditions, chemical influences, life form, hybridity, breeding system, karyotype, chromosome size, and specific genes. The relation between several of these factors and polyploidy in *Gilia* and *Ipomopsis* has been discussed in previous papers (Grant 1952c, 1953, 1956b). The apparent irrelevance of one additional factor will be briefly mentioned here. The rule that the frequency of polyploids increases in higher latitudes and colder climates does not appear to hold true in the Polemoniceae. Indeed, the main centers of polyploidy in this family appear to lie in the tropical and desert floras.

The differences between genera and sections of the Polemoniaceae as to the frequency and the level of polyploidy are probably to be sought in differences in the complex interplay of the various individual factors cited above and perhaps other as yet unidentified factors. We can analyze the effects of single factors and simple combinations of factors in the promotion of polyploidy. Some initial attempts have already been made in this direction for the Polemoniaceae, as noted above. But we obviously do not know enough yet to assess the combined effect of a multiplicity of variables on polyploidy. A completely satisfying explanation of the group-to-group differences in the evolutionary role of polyploidy still eludes us.

Chromosome number is only one of several visible features of the karyotype possessing taxonomic, genetic and evolutionary significance. Flory (1934, 1937) has investigated the position of the centromere in the somatic chromosomes of several selected species. His principal findings are summarized in Table 11. A species of the primitive genus *Polemonium* is seen to have a more symmetrical karyotype than its more advanced relatives in *Collomia* and *Phlox*. The *Gilia* species examined are likewise more symmetrical in their karyotypes than the highly derived *Linanthus* species. A phylogenetic trend toward increasing asymmetry in the karyotype, such as has been found in *Crepis* and other plant groups, is suggested by these data. However, much more work needs to be done on chromosome morphology in the Phlox family before any generalizations can be made in this field.

With respect to the size of the chromosomes, the following qualitative relationships prevail. Minute chromosomes with a large nuclear-cytoplasmic ratio are found in the tropical genera *Cobaea*, *Cantua* and *Loeselia*. Small chromosomes characterize *Linanthus* and *Leptodactylon*, and medium-small chromosomes *Collomia*, *Phlox*, *Gilia*, *Navarretia* and *Eriastrum*. Large chromosomes presenting a swollen aspect in meiotic cells occur in *Polemonium* and *Ipomopsis*. Some of these size differences can be seen in Figure 62. Flory (1937) gives the following measurements for the length of the longest chromosomes in the somatic complements of eight representative species: *Polemonium boreale*, 11.5 mu; *Ipomopsis rubra*, 10 mu; *Collomia grandiflora*, *Phlox bifida* and *Gilia achilleaefolia*, 9 mu; *Eriastrum virgatum*, 7.5 mu; *Linanthus dianthiflorus*, 4.5 mu; *Cobeaa scandens*, 2.5 mu.

Polyploid species, at least in *Gilia*, usually have smaller individual chromosomes than the related diploids, with the result that the nuclear-cytoplasmic ratio does not vary significantly in a polyploid series.

The differences between minute and normal-sized chromosomes happen to coincide with tribal boundaries within the family. The presence of minute chromosomes in the tropical woody genera is in line with the idea first suggested by Darlington (1937: 84) that the conditions for mitosis in cambial cells of woody plants set a limitation on chromosome size. The temperate herbaceous genera are obviously not subject to this limiting factor. Among the temperate

Table 11. Position of the Centromere in Representative Members of the Phlox Family (from Flory)

	No. of chromosome pairs with centromere:				N
	Median	Sub-median	Sub-terminal	Terminal	
Polemonium Tribe					
Polemonium boreale	8	—	1	—	9
Collomia grandiflora	—	1	5	2	8
Phlox bifida	2	—	5	—	7
Gilia Tribe					
Gilia capitata	5	—	4	—	9
Gilia achilleaefolia	2	3	4	—	9
Ipomopsis rubra	—	2	5	—	7
Eriastrum virgatum	1	2	4	—	7
Linanthus dianthi-					
florus	—	3	5	1	'9

Polemoniaceae a trend is apparent from large to small (but not minute) chromosomes in each of the two tribes. In the Polemonium tribe this trend is seen in comparisons of *Polemonium* with *Phlox*, and in the Gilia tribe in comparisons of *Ipomopsis* and *Linanthus*. Similar phylogenetic trends from large to small chromosomes are found in other herbaceous plant groups, as for example *Crepis* and its relatives.

CHAPTER VII

PHYLOGENY

The term phylogeny is currently being used for two different disciplines, the tracing of the history of a group from its fossil record, and the reconstruction of a hypothetical evolutionary history from the comparative study of recent forms. So different are the methodologies of the two approaches that it has been questioned whether the same term phylogeny should be used for both (Simpson 1953: 214). Nevertheless the ultimate purpose of the investigation is the same in each case and the differences of methodology are determined by the nature and limitations of the evidence rather than by the philosophy underlying the investigation. No comparative morphologist would fail to welcome the evidence contributed by a fossil record. If such a record is lacking, scientific curiosity and the speculative urge being what they are, the worker will proceed on the basis of comparisons between recent forms alone.

Fortunately this approach is not wholly unfruitful. This is due largely to the fact that evolutionary rates vary widely within a given group, some lines changing relatively little from the ancestral condition while other lines are undergoing rapid transformations. Furthermore, the greater wealth of recent forms and the better state of their preservation compensates the comparative morphologist to some degree for the lack of direct fossil evidence concerning their past. The greatest difficulty, of course, lies in the finding of reliable criteria for distinguishing between what is primitive and what is advanced within the group under consideration. (For an illuminating discussion cf. Sporne 1956).

We have one fossil record of the Polemoniaceae.[1] Fossil fruits of a *Phlox* probably close to *P. sibirica* or *P. borealis* have been found in a Pleistocene deposit from Fairbanks, Alaska (Chaney and Mason 1936). All that this really tells us is that Phloxes, which now inhabit the area around Fairbanks, Alaska,

---

[1] Since this chapter was written a second fossil record has been reported. Pollen of *Polemonium* has been found in Pleistocene deposits in Great Britain (Godwin 1956).

were also there several thousand years ago. As far as the Polemoniaceae are concerned, we have our choice at present between an uncertain and speculative phylogeny based on comparative morphology or no phylogeny at all.

THE PRIMITIVE CONDITION

In the Polemonium tribe, as shown in Table 3, the characteristics of the various genera fall into a graded series. One extreme in the series is represented by rhizomatous perennials of mesic habitats with alternate pinnate leaves, a herbaceous calyx, campanulate corolla, equal stamens, many-seeded capsules, and 9 pairs of chromosomes (*Polemonium*). At the other ends of the series are perennial or annual herbs with opposite linear leaves, membranous calyx-tubes, salverform corollas, irregularly inserted stamens, one-seeded capsule locules, and 7 pairs of chromosomes (*Phlox, Microsteris*); or leafless desert ephemerals with a scarious calyx, sessile stamens, minute autogamous flowers, and 6 pairs of chromosomes (*Gymnosteris*). The extremes are connected by transitional stages as represented by *Collomia*.

Can we establish which end of the series represents the primitive and which end the advanced condition? Three considerations point convincingly to *Polemonium* as the bearer of the most primitive characters in the tribe. In the first place, the perennial habit, alternate phyllotaxy, open corollas, equal stamens, and many-seeded fruits approach more closely than any alternative characteristics in the tribe the condition found in the most primitive living Dicotyledons, the woody Ranales. In the second place, through *Polemonium*, and through it alone, the tribe can be related with other, tropical genera of the family, whereas the more specialized genera in the tribe are related through *Collomia* with *Polemonium*.

Finally, the derivation of an annual from a perennial life cycle, of opposite from alternate leaves, of a histologically differentiated from an undifferentiated calyx, of salverform or autogamous flowers from open campanulate ones, of an irregular from a regular androecium, and of indehiscent one-seeded fruits from dehiscent many-seeded ones, can easily be explained in terms of known processes of morphological reduction, whereas the reverse evolutionary trends from the more specialized to the more generalized condition would be very difficult to account for. In the light of these considerations it is reasonable to look upon the Polemonieae as a reduction series running from *Polemonium* through *Collomia* to *Phlox* with numerous side branches leading to the minor genera *Allophyllum, Gymnosteris* and *Microsteris*.

Among the temperate genera of the Phlox family, only *Polemonium* approaches *Cobaea*. The two genera are alike in perennial habit, mesic ecology, pinnate leaves, herbaceous calyces, often large and campanulate corollas, and low point of stamen insertion. In certain species of *Polemonium* the seeds possess very narrow bordering fringes which are perhaps vestiges of the wings on the seeds of *Cobaea* and other tropical genera. Geographically *Polemonium* is well developed in the mountains of Mexico where some morphologically unadvanced species such as *P. grandiflorum* occur. In fact, *Polemonium* is the only temperate genus, with the exception of one species each of *Ipomopsis* and *Phlox*, which does range deep into Mexico. *Cobaea*, on the other hand, overlaps broadly with *Polemonium* in the Mexican highland, reaching its northern limit with *C. pringlei* in the Sierra Madre near Monterrey. Both the morphological and the distributional evidence speak in favor of a relationship between *Polemonium* and *Cobaea*.

In the Gilia tribe a number of reduction series are also evident, as shown in Table 5. One end of the series is represented by subshrubs or perennial herbs with alternate leaves, regular calyx and rotate or tubular corollas, such as occur in *Ipomopsis*, *Eriastrum*, and *Gilia* sect. *Giliastrum*. At the other ends of the series are annuals with opposite leaves and salverform corollas (*Linanthus* sect. *Leptosiphon*); desert annuals with bristle-tipped leaves and bilabiate corollas (*Langloisia*); and inhabitants of vernal pools with zygomorphic calyces, 2-lobed stigmas and indehiscent capsules (*Navarretia* sect. *Navarretia*).

This series, like that in the Polemonium tribe, only makes sense when read in one direction. It is highly probable that the alternate-leaved, regular-flowered shrubby forms are the most primitive in the Gilieae. These characteristics, and not the alternative ones found in the tribe, approach the condition exemplified by the most primitive Dicotyledons. Furthermore the alternate-leaved shrubby Gilieae can be related to other phylads outside the Gilia tribe, whereas the highly specialized annual types are related through various intermediate forms with the perennial species. Finally, the change from a perennial to an ephemeral life cycle, from alternate and pinnate leaves to opposite or bristle-tipped or linear ones, and from open campanulate flowers to zygomorphic or salverform flowers is more likely to occur, especially as a long-term trend, than the reverse change.

The predominantly herbaceous genera *Ipomopsis* and *Eriastrum* and the predominantly herbaceous phylad *Leptodactylon-Linanthus* all include some shrubby species, which are regarded as primitive within their respective lines on the assumption that the prevailing evolutionary trend is from woody plants to herbs. Furthermore the shrubby species in question, *Ipomopsis multiflora*, *I. pinnata* and *I. gloriosa*, *Eriastrum densifolium*, and *Leptodactylon*, have enough resemblances among themselves to suggest that they are phylogenetically more closely related to one another than are their respective herbaceous derivatives.

By subtracting the obviously specialized features from each species –

the salverform corolla shape from *Leptodactylon californicum*, the sinuous corolla tube from *Ipomopsis multiflora*, the reduced chromosome number in *Ipomopsis* and *Eriastrum*, etc. – we arrive at a composite picture of a hypothetical ancestral form for this tribe. The progenitor should be a shrub with alternate pinnately dissected leaves, a regular calyx, large regular tubular corollas, oval erect anthers, and 9 pairs of chromosomes. *Ipomopsis gloriosa* from Baja California conforms closely enough to this description in most respects to afford a concrete living illustration of the kind of common progenitor from which the other Ipomopsises, Eriastrums, Leptodactylons and their annual relatives might have evolved.

The genus *Gilia* has its most primitive forms in the perennial, broad-leaved plants of the section *Giliastrum* with their campanulate corollas and low point of stamen insertion. *Gilia rigidula* exemplifies (but does not necessarily represent) a species from which the rest of the genus could be derived. Through it the genus *Gilia* approaches most closely the shrubby members of the other alliances in the tribe.

Though the shrubby Gilieae are separated by a wide gap from other elements in the family, a gap which is no doubt the result of extinction of the intermediate forms, this gap does narrow down significantly in one direction. The only genus outside the Gilieae which exhibits evidences of particular relationships to this tribe is *Loeselia*. This genus, though very different as to leaf form, seeds and corolla symmetry, has calyces of the same type as occur throughout the Gilieae and is adapted to seasonally dry situations as are the Gilieae. The most likely hypothesis which we can offer in the present state of our information is that the Gilia tribe developed out of some unknown and probably long extinct common ancestor in or near *Loeselia*.

The temperate tribes of Polemoniaceae thus converge upon different tropical forms, the Polemonieae upon *Cobaea* and the Gilieae upon *Loeselia*. Now the tropical Polemoniaceae are interrelated among themselves. *Loeselia* is related to *Bonplandia* which in turn is related to *Cobaea* and *Cantua*.

It is among the tropical tribes that the greatest number of primitive morphological features in the family are found. These primitive characters are not concentrated in any single genus, but are irregularly distributed among the existing tropical genera.

Thus the primitive life form of a small tree or large shrub is represented in *Cantua* and *Bonplandia*. Broad, alternate leaves not reduced or especially differentiated in the inflorescence are found in *Cobaea, Cantua, Bonplandia*, and part of *Loeselia*. A herbaceous calyx composed of sepals fused only at the base occurs in *Cobaea*. A moderately large tubular or campanulate corolla ending in five equal oval lobes is found in *Cobaea, Cantua* and *Huthia*. Equal, non-epipetalous stamens occur in *Cobaea* and *Cantua*; in these genera also an extra vestigial whorl of stamen traces has been found. A septicidal manner of capsule dehiscence along the lines of union of the carpels is preserved in *Cobaea*. Nine pairs of chromosomes are found in *Loeselia*. The tropical ecology itself, in one or another of its aspects, is ancient and probably original for the Dicotyledons as a whole and for the Polemoniaceae in particular.

Several morphological characters occur so generally among the diverse tropical genera of Polemoniaceae as to suggest that they are primitive for this family. The characters in question may not be primitive in relation to the long-term trends of Dicotyledon evolution; yet they may represent an ancestral condition within the Polemoniaceae. As examples of such characters we may mention pinnate lobing of leaves, winged seeds, colorless embryos with non-foliaceous cotyledons, and minute chromosomes.

As previously noted, no single species or genus of tropical Polemoniaceae has retained all of the primitive features found within the family. Each living form presents certain definite specializations. The vine-like habit, the leaf tendrils, the gigantism of the flowers, and the polyploid constitution of *Cobaea* are probably specializations which have developed during the evolutionary divergence of this genus from the unknown and extinct ancestral Polemoniaceae. The synsepalous calyx, loculicidal capsule, and high polyploidy of *Cantua* represent specializations attained in another line. The zygomorphic calyx and corolla, the salverform corolla, the epipetalous stamens, and the small-sized capsule of *Bonplandia* are still other specializations. *Huthia* and *Loeselia* diverge from primitiveness in their synsepalous calyx, epipetalous stamens, and loculicidal capsule; in addition, *Huthia* possesses an apparent modification in its linear leaf form; while in *Loeselia* the zygomorphic, partly membranous calyx of all species and the herbaceous habit and salverform corolla of some species are evident specializations.

Each living tropical genus then comprises a different mosaic of primitive and advanced features. This observation, together with two others, the prominent morphological gaps between the genera, and their relatively poor development into species[2], indicates that the tropical genera of Polemoniaceae are an assortment of relict groups, each one of which has followed its own particular course of specialization. Considered as a whole, they afford us several glimpses, but no single coherent picture, of the primitive condition of the family.

PARALLELISM IN EVOLUTION

The difficulty of generic delimitation in the Polemoniaceae has been commented on by numerous authors. It is fair to inquire into the causes responsible

[2] The size of *Cobaea*, the only large tropical genus in the family, may be partly illusory, since some of the taxonomic species recognized at present may have to be reduced to subspecies or minor variants when the genus is better understood.

for the situation in this family. Three factors can be suggested: (i) lack of knowledge; (ii) parallelism in evolution; and (iii) a reticulate pattern of relationships resulting from the hybrid origin of some of the lines.

The earliest students of the Polemoniaceae, Bentham, Peter, Brand and others, did not know the plants outside of the herbarium or garden, and Gray had only a passing acquaintance of them in the field. Furthermore the representation of the family in herbaria during the 19th century was inadequate by modern standards. The lack of knowledge about the plants should not be overemphasized, however, since with the evidence then available it was possible to work out the taxonomy of the tropical and subtropical genera in a generally satisfactory way. Brand's classification of this part of the family in 1907 has stood the test of time. The only major changes considered necessary in the present study were the relinquishment of Brand's subfamilies as natural units and the transfer of the tribe Bonplandieae from one main subdivision to another.

Considering that the tropical and subtropical genera are still rather poorly known in the wild and in the garden, it is significant that the classical students of the Polemoniaceae could achieve so much better results here than in the temperate genera. Some of the most familiar North American groups, on the other hand, such as the showy and much collected *Ipomopsis aggregata* group and the cespitose Phloxes, are still today in a very confused state. Evidently, therefore, the taxonomic difficulties in the Polemoniaceae are not due entirely to a lack of information about the plants, but are to a certain extent inherent in the evolutionary pattern of the family, or more properly one part of the family, the temperate genera.

A most important stumbling-block in the way of the recognition of phylogenetic relationships in this part of the family has been the parallel evolution of similar morphological and ecological traits in separate phylads. To mention a single example of parallelism, the trend from perennial to annual life cycle has gone on repeatedly in nearly every temperate genus, only *Leptodactylon* being wholly perennial and only a few derived and mostly small genera being wholly annual.

Amongst the annuals, moreover, there are invariably some species which have changed over from outcrossing to selfing in the breeding system and which have highly reduced, small flowers. Correlated with this transformation of the breeding system, at least in the genus *Gilia*, is an increase in the frequency of ring bivalents in pollen mother cells. Another character which seems to be correlated with the annual habit is the mucilaginous nature of the seed coat when wet. Thus in the tribe Polemonieae there are very few perennial species which do not have non-mucilaginous seeds and very few annuals which do not have mucilaginous seeds. There are more exceptions to this rule in the tribe Gilieae, though the correlation is evident here too, nearly all of the annuals but also many of the perennials having mucilaginous seed coats.

By merely considering uncritically a large number of characters – annual habit, small flowers, included stamens and styles, high proportion of ring bivalents, and mucilaginous seeds – a taxonomist may be led completely astray in grouping together the reduced end products of several independent phyletic lines. The taxonomic history of the family shows that just this mistake has been made repeatedly. For instance, Gray proposed a special section (*Courtoisia*) in order to group together *Collomia heterophylla, Allophyllum glutinosum* and relatives, and *Gilia capillaris*. Milliken made another grouping (under subgenus *Microsteria*) of *Microsteris gracilis, Ipomopsis depressa*, and *Allophyllum*. Brand created the section *Phlogastrum* for *Microsteris gracilis, Ipomopsis minutiflora, Gilia tenerrima* and *Allophyllum*. Mason and A. Grant brought together in their subgenus *Kelloggia* three species of *Gilia* sect. *Saltugilia* and one species of *Ipomopsis*.

The species thus united are in all cases reduced annuals and in the majority of cases self-pollinators. Their true relationships, revealed by other characters definitely allying them to certain species in one phylad or another, are obscured by evolutionary convergences.

Other characters showing parallelisms are the indehiscent few-seeded capsules in some species of *Gilia* and *Navarretia*; the irregular stamen insertion in *Gilia, Navarretia* and *Collomia*; the zygomorphic corollas in *Ipomopsis* and *Langloisia*; the membranous calyx sinuses in *Phlox* and the Gilia tribe; the opposite leaves and salverform corollas in *Phlox* and *Linanthus*, representing advanced forms in two distinct tribes; and the reduction of chromosome number from n = 9 to n = 7 or 6 independently in the two tribes. That parallelism is involved is indicated by the fact that some relatives of each line exhibit the more primitive condition for each character, such as dehiscent many-seeded capsules, regular corollas, alternate leaves, etc.

As far as I can see the consideration of parallelisms is not encouraging to the prospects of purely mechanical methods in systematics, such as the punching of cards and their classification by IBM machines, as advocated by several authors. If the more obvious characters are selected for scoring we can expect a statistical confirmation of Gray's and Brand's groupings of *Microsteris gracilis, Allophyllum gilioides, Collomia heterophylla*, and certain *Gilia* and *Ipomopsis* species. Convenience is apt to go hand in hand with artificiality in the classification of complex groups. If the family has to be analyzed along phylogenetic lines first in order to discover the characters not affected by convergence, the punched card method will have saved the investigator none of the really critical phases of the work.

Another and more time-honored method of systematics which is invalidated by the recognition of widespread parallelisms is the emphasis on key characters. Greene, whose work revealed a profound understanding of the Polemoniaceae, was nevertheless forever declaiming on the importance of this or that character of supposedly generic value. In 1887 it was the calyx, in 1898 the seed coat.

"The one general conclusion reached by me, after eighteen years of field experience with these plants [Polemoniaceae] is, that characters of form of corolla and length, insertion, direction, etc. of stamens may be set aside as wholly incompetent to furnish means of defining genera; and that, by the calyx alone, especially as it appears not in flower, but in its after developments, in and of itself and in its relation to the fruit, we may limit and define good acceptable genera, made up of plants agreeing in habit, and in some other minor points. In order to make use of the characters indicated, we must, I think, entirely lay aside, what I conceive to be a mere prejudice, the notion that the form of the corolla, and the direction of the stamens – whether erect and straight, or curved and declinate – need to be considered at all, in the matter of generic diagnosis" (Greene 1887).

"At present I am disposed to adopt it as a principle that species with mucilaginous seeds are nowhere, in this family, to be placed as congeneric with such as have seeds devoid of the gummiferous coating" (Greene 1898).

Curiously enough the same doctrinaire approach to taxonomic characters which led Greene to split genera caused Gray to reduce nearly all of them to *Gilia*. "The genera are very difficult to define. Even *Phlox* has a species with prevailingly alternate leaves and four or five ovules in each cell. The character of unequally inserted stamens, by which to distinguish *Collomia*, breaks down completely . . . So it has become necessary to incorporate *Collomia* in *Gilia*.." (Gray 1886). "*Gilia* is certainly a polymorphous as well as a large genus; but definite characters are vainly sought for dividing it and for keeping *Navarretia* separate" (Gray 1870).

The phylogenetic approach to taxonomy, as opposed to the typological approach, emphasizes relationships rather than characters per se in the grouping of taxa. This calls for more detailed descriptions and the recognition of exceptions within and convergences between the various groups. If all that is desired is a pigeon-hole classification the extra labor involved is probably not justified; if the goal is a natural classification the extra labor is indispensable.

THE INFLUENCE OF HYBRIDIZATION ON PHYLOGENY

In five sections or comparia of the Polemoniaceae genetic relationships of species have now been studied by biosystematic methods. These are the small genus *Allophyllum*, *Ipomopsis* sect. *Ipomopsis*, and three sections of

Gilia, namely the sections *Gilia*, *Saltugilia*, and *Arachnion*. Natural hybridization has been found to occur in all of these groups and has strongly influenced the course of evolution of four of them (Grant 1953, Grant and Grant 1955, 1956; see Fig. 66). Cases of introgressive hybridization have also been described in *Phlox* and *Polemonium* (Anderson and Gage 1952, Braun 1956).

It would not be justified to conclude from this that four-fifths of the sections in the temperate genera of Polemoniaceae have been affected in their development by hybridization, since the foregoing sample is a strongly biased one. The groups selected for intensive study were chosen partly because they did show the effects of natural hybridization, there being a special interest in this process for its own sake, and partly because the groups standing in greatest need of taxanomic attention are those in which hybridization has had strong effects. The taxonomically simpler and non-mongrelized groups were worked out more or less satisfactorily by earlier generations of botanists.

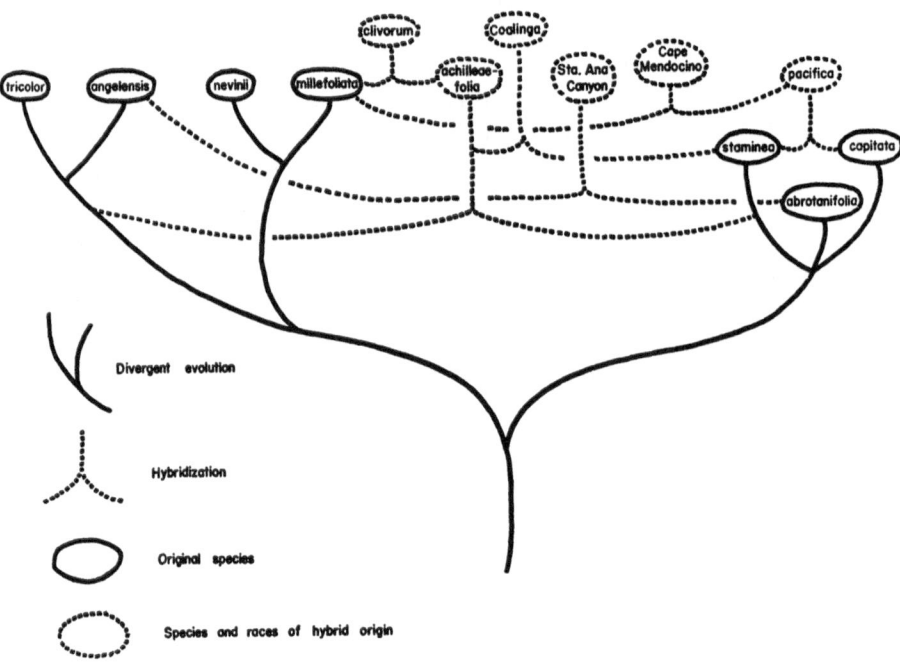

Fig. 66. Phylogenetic web of the Leafy-stemmed Gilias. (from Grant.)

Nevertheless there can be no doubt that the worst-tangled comparia in the entire family, such as the Leafy-stemmed Gilias, the *Ipomopsis aggregata* group, the Allophyllums, the cespitose Phloxes, and above all the Cobwebby Gilias, are hybrid complexes. More specifically they are homogamic complexes and polyploid complexes in which the variability of the species has been enhanced and the morphological and ecological distinctions between the species more or less obliterated by a long history of hybridization on the diploid and polyploid levels.

Grant has proposed to use the concept of the homogamic complex as a model of the pattern of evolution in the higher categories of angiosperms. "In homogamic complexes the hybrid derivatives are diploid, sexual populations with a normal chromosome cycle at meiosis, rather than apomicts, polyploids, permanent structural heterozygotes, or clones. There is consequently much less restriction on gene recombination in the derivatives of homogamic complexes than in the end products of agamic, polyploid, heterogamic or clonal complexes. As a result, homogamic complexes possess an open system of evolution, as opposed to the closed system of evolution which characterizes the other types of hybrid complex. . . . [In a homogamic complex] hybridization could continue indefinitely without reducing the ability of the group to make progressive evolutionary advances.

"The Angiosperms as a whole have undergone many progressive evolutionary changes while developing as a phylogenetic net rather than as a conventional phylogenetic tree. In the light of independent evidence for a widespread occurrence of hybridization in the past history of the Angiosperms, this combination of a reticulate pattern of relationships with progressive trends of specialization suggests that the class may have evolved as a succession of homogamic complexes" (Grant 1953).

Reticulate evolution due to the combined operation of speciation and hybridization represents a basic evolutionary pattern. It is as basic, for example, as the quantum evolution of Simpson, and has the additional advantage of resting on experimentally verified foundations. If we are to base our ideas of macroevolution on inductive methods of inquiry we will have to give up, in many plant groups, the old symbolism of the phylogenetic tree in favor of the symbolism of a phylogenetic net (Fig. 66).

The concept of reticulate evolution can be profitably applied to the phylogeny of the temperate Polemoniaceae. If races and species of hybrid origin can confuse the taxonomy of a section, then sections derived from species of hybrid origin can confuse the taxonomy of a genus, and genera descended from interspecific hybrids can confuse the classification of a tribe. To make matters worse, the ancient reticulations may never be identified as such. The phylogeny of the Polemoniaceae in the past, judging from the situation in certain experimentally analyzed recent groups, probably consists of a mixture of dichotomous branches and web-like anastomoses.

The two extreme forms in the genus *Allophyllum* are *A. glutinosum* and *A. divaricatum*. The former is characterized by a mesic ecology; many-lobed

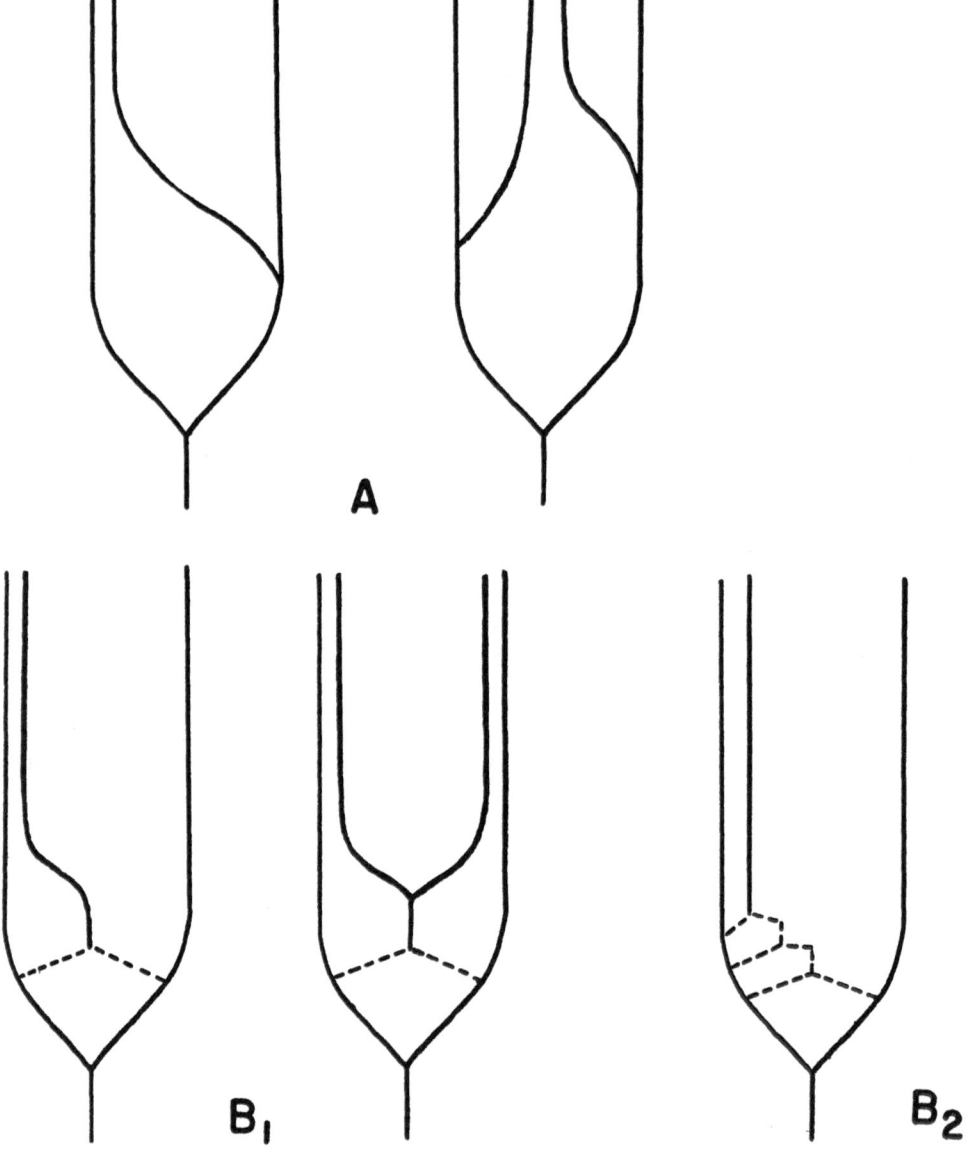

Fig. 67. Two ways by which parallelisms can develop in evolution. Parallel selection is assumed in each case.
a) Independent parallel mutations at homologous loci. *b*) Hybridization followed by segregation in the direction of one or both parental species (b_1), or followed by backcrossing, viz. introgression (b_2).

pinnate leaves; bluish, open, funnelform corollas; long, declined, nearly equal stamens; multi-seeded capsules; and 9 pairs of chromosomes. *Allophyllum divaricatum* has a more xeric ecology; simpler leaves; a long, slender, almost salverform corolla with a reddish tube; unequally inserted, very unequal, long stamens; one-seeded capsule cells; and 8 pairs of chromosomes. The foregoing differences are very interesting since they are the differences between *Polemonium* and *Collomia*. In fact if only *A. divaricatum* existed there would be a strong temptation to include it in the genus *Collomia*, whereas the resemblance of *A. glutinosum* to the genus *Polemonium* is unmistakable.

Is this a case of the origin of a genus (*Allophyllum*) from some ancient hybrid between members of two generic lines (*Polemonium* and *Collomia*) when their divergence had not yet passed the limits of crossability? Or is it an extreme example of the phenomenon of parallelism already referred to? The very fact that there does not appear to be any way of distinguishing between these two possible explanations may be one of the most significant features about this case.

Parallel evolution is a resultant of the combined action of parallel genetic variation and parallel selection. The latter force is easy enough to account for. Different phyletic lines inhabiting the same environment will be exposed to some of the same selective factors.

Parallel variation, on the other hand, can arise either through independent mutations at homologous loci in the different phylads, or through the sharing of the same mutations by separate phylads as a result of hybridization and gene exchange. The alternative modes of origin of parallelisms are illustrated diagramatically in Figure 67. What we know about both mutation and hybridization suggests that it is genetically easier in many plant groups for parallel variations to arise by hybridization than by independent mutations.

The diploid state permits organic populations to maintain a storehouse of mutations. The key to this storehouse is sexual reproduction. Now, hybridization between species represents an extension on the grand scale of the possibilities of sexual reproduction, by opening up the storehouse of mutations to members of distinct species (Grant 1950: 296–297). Hybridization renders independent mutation superfluous as a source of parallel variations in separate phylads.

The ability to segregate and vary in the direction of the ancestral species is a known characteristic of interspecific hybrids and their derivatives. This ability probably forms the basis for the evolutionary development of many parallelisms.

CHAPTER VIII

PHYTOGEOGRAPHY

In the present chapter we shall use our taxonomic knowledge of the Phlox family as the basis for an analysis of phytogeographic patterns. Clearly the phytogeographic edifice can be only as strong as its taxonomic foundations. We should know both the strength and the weakness of this foundation so that we can evaluate the reliability of the phytogeographic conclusions.

Nothing can be more certain than that future taxonomic studies in the Phlox family will result in an alteration of many species lines and many distribution areas. *Langloisia*, for example, may prove to consist of four rather than five species. The species lines in *Cobaea* may have to be drawn quite differently when this genus is better known. The European and Asiatic Polemoniums are regarded in the present conservative treatment as falling into four species, whereas a much larger number of species is recognized by Russian botanists, and the true complexity of the genus in the Old World may lie somewhere between these contrasting estimates. These and other questions have been touched upon in Chapter 3.

The distribution areas of many species may be expected to undergo changes corresponding to different taxonomic interpretations. Thus *Phlox sibirica* was formerly regarded as occurring both in Asia and the American Arctic. Wherry in his recent monograph of *Phlox* segregated the old *P. sibirica* into separate Asiatic and North American branches, calling the former *P. sibirica* proper and the latter *P. borealis*. I have followed Wherry, though not without misgivings, and it may yet turn out that *Phlox sibirica* occurs as a polytypic species on both sides of Bering Straits. Similar distributional problems will be found in every other large genus in the family.

Any census of the number of species of Polemoniaceae in a geographical area which we can carry out today will consequently be subject to revision in the light of future taxonomic researches. The same is true of estimates of the frequency of endemic species in a region, or of the number of geographical areas occupied by individual species, or of the overall composition of the polemoniaceous flora in different parts of the world. In matters of detail the present phytogeograph'cal picture can have no finality.

Nevertheless, the Phlox family is now well enough understood to justify certain broad general conclusions concerning the geographical relationships. Even supposing that there are eight or ten rather than four species of Polemonium in Eurasia, our general conclusion that the polemoniaceous flora of that continent is small will still hold. The disposition of *Phlox sibirica* will decide whether the number of species of Polemoniaceae common to Eurasia and North America is two or three, but in terms of the broad picture this is again only a minor detail. It is doubtful whether future taxonomic studies in the Phlox family will affect the main conclusions reached in this chapter in any major way.

CENTERS OF DISTRIBUTION

The Polemoniaceae occur as native plants on three continents, North America, South America and Eurasia, and as weeds on two additional continents, Africa and Australia. The natural distribution of the family as understood at the present time is shown in Figure 68.

The outstanding feature of this world distribution is its departure from evenness of representation in the different areas. Of the 316 species of Polemoniaceae, 290 occur in North America, 26 occur in South America, and only five inhabit the vast area of Eurasia. Three species are common to North and South America (*Loeseli? glandulosa, Polemonium micranthum, Microsteris gracilis*), and two species are common to North America and Eurasia (*Polemonium caeruleum* and *P. boreale*).

On the North American continent similar contrasts are found in the richness of the polemoniaceous flo ₁ of different regions. Thus Alaska and New England have six species of Pole¯ioniaceae each, the states of North Dakota and Florida have seven each, Kansas fourteen, and the whole eastern half of the continent from the 100th meridian to the Atlantic coast has twenty-six species. By contrast, California with 161 species possesses over half the species in the family. The main concentration of species in the Phlox family evidently lies in western North America.

It was desirable to establish the centers of distribution for the Phlox family with greater accuracy than could be achieved by spot-checking the floras in various political-geographical areas. The first step towards a more analytical investigation was the drawing up of distribution maps for each species in the family. These maps were compiled from herbarium sources and published records. Some basis of classification was next needed for reducing the distributional data to a manageable form. A satisfactory basis for classifying the data would consist of a subdivision of the world area of the Polemoniaceae

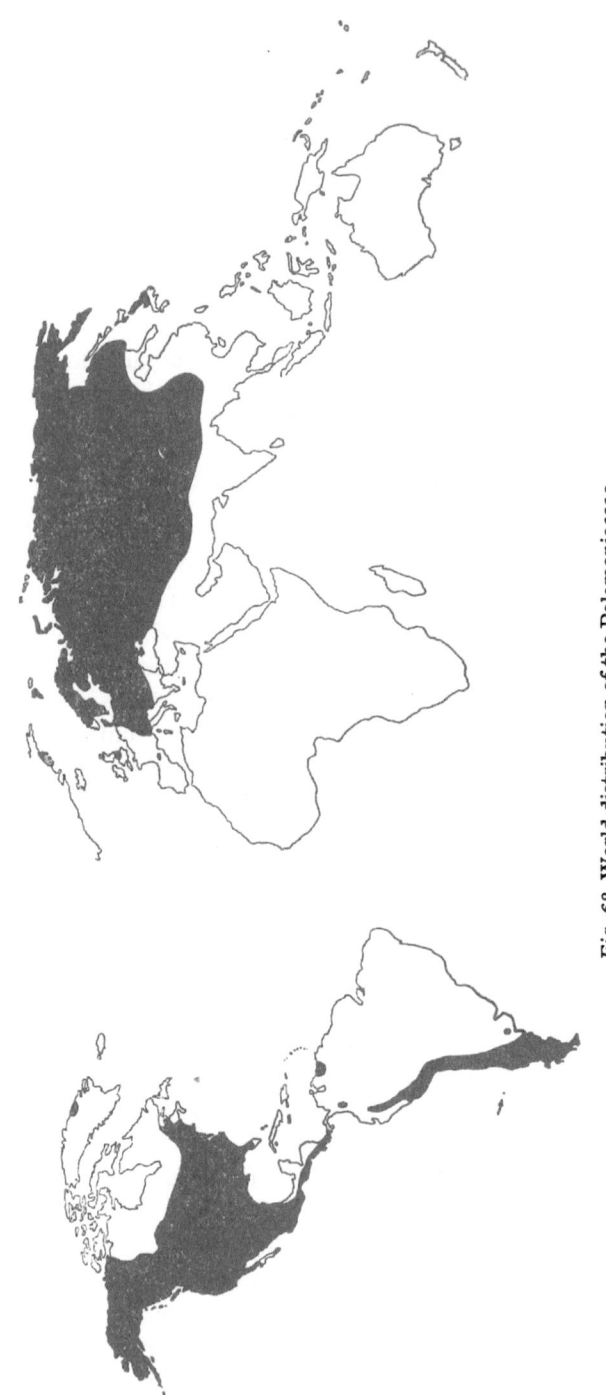

Fig. 68. World distribution of the Polemoniaceae.

shown in Figure 68 into smaller natural areas within which a significant number of the species reach some or all of their limits of distribution.

Biotic classifications of North America and the world have been proposed by Harshberger, Schantz and Zon, Clements, Dice, Schmidt and other authors. Attempts to fit the distribution areas of the Polemoniaceae into the available categories did not, however, meet with much success. In some cases the distribution areas of many species of Polemoniaceae cut across several biotic provinces. In other cases the recognized biotic provinces are drawn too broadly.

It appears that the biotic regions recognized by most biogeographers are delimited on the basis of the distribution of dominant vegetation types. One does not have to carry out very much field work on the Polemoniaceae to realize that the species of this family do not coincide in their distributions with the species of oak, pine, grass and scrub which define the major biotic provinces. What we needed was a biotic classification of the world based, not on forest trees, grasslands and desert shrubs, but on the Polemoniaceae themselves.

A set of natural areas applicable to the Polemoniaceae was therefore drawn up from the distribution maps of the Polemoniaceae. The areas were blocked out by the process of repeatedly scrutinizing the maps until certain boundary lines, representing the extreme geographical limits of a fair number of the indigenous species, became apparent. The areas thus delimited were next compared with the major physiographic provinces as given by Atwood (1940). In several cases this exercise suggested shifts in the boundary lines which brought the natural polemoniaceous areas into correspondence with natural physiographic areas.

The system of natural areas adopted for the purpose of our present discussion is by no means the only possible system. In fact, I have modified the classification during the course of the analysis whenever a practical advantage was gained by the change. Thus temperate South America could be treated as a single area, and was at first so treated. But within this large area the polemoniaceous floras of the Andean highlands and of the lowlands are with rare exceptions specifically and even generically distinct. It seemed most useful to recognize two areas in temperate South America.

Cismontane California, again, was at first treated as a single area. It later became apparent that two floristic areas, a northern and a southern area, are involved. Of 130 species of Polemoniaceae indigenous to cismontane California only 21% are common to both the northern and southern divisions. The parts of California lying west of the deserts have consequently been repartitioned for our purposes into two natural areas.

The natural areas arrived at by the foregoing empirical procedure are named in the left-hand column of Table 12 and mapped in Figures 69 and 70. An analysis of the distributional data for the Phlox family was now carried out within the framework of this geographical classification. For each natural area three botanical characteristics were determined. It was of interest, in the first place, to determine the number of species of Polemoniaceae occurring in each area; secondly, to determine the number of species endemic within each region; and finally, to describe the floristic composition of the different natural areas.

The species density and percentage of endemism in the different areas are summarized in the right-hand columns of Table 12 and shown graphically in Figures 69 and 70. The salient facts are as follows.

A. *Central America and Venezuela*. Three genera are found here, *Cobaea*, *Bonplandia* and *Loeselia*. One section of *Cobaea* and one section of *Loeselia* are endemic within the area. The most important genus, *Cobaea*, is represented by 13 species, all of which are endemic. The southernmost limits of *Loeselia* and *Bonplandia* lie within this area. The frequency of endemics is 15/22 or 68%.

B. *Andes*. Three tribes are represented in the Andean highland, the Cobaea, Cantua and Gilia tribes. The Cantua tribe, comprising two genera, is endemic

Table 12. Richness and Endemism of the
Polemoniaceous Flora in different Natural Areas

Natural Area	Number of Species						Endemism	
	Cobaea Tribe	Cantua Tribe	Bonplandia Tribe	Polemonium Tribe	Gilia Tribe	Total	No. Species	%
A Central America and Venezuela	13	—	9	—	—	22	15	68
B Andes	3	8	—	—	4	15	13	87
C Temperate South American Lowlands	—	—	—	4	6	10	6	60
D Mexican Highland	3	—	9	6	9	27	11	41
E Northern Sierra Madre	—	—	1	14	24	39	4	10
F Colorado Plateau	—	—	—	23	37	60	10	17
G Sonoran Desert	—	—	—	—	16	16	2	12
H Mojave Desert	—	—	—	2	39	41	11	27
I Great Basin	—	—	—	34	39	73	9	12
J Cismontane Southern California	—	—	—	12	42	54	19	35
K Northern California	—	—	—	27	71	98	43	44
L Pacific Northwest	—	—	—	24	16	40	4	10
M Alaska and Northwestern Canada	—	—	—	6	—	6	2	33
N Rocky Mountains	—	—	—	35	23	58	7	12
O Great Plains	—	—	—	9	11	20	2	10
P Eastern North America	—	—	—	24	2	26	20	77
Q Eurasia	—	—	—	5	—	5	3	60

here. The genus *Cobaea* reaches its southern limit within the area. The percentage of endemism at the species level is high, 87%.

C. *Temperate South American Lowlands.* The Gilia and Polemonium tribes are represented here by a total of 10 species, six of which are endemic.

D. *Mexican Highland.* This is the only area in which four tribes occur together. The genera present are *Cobaea, Bonplandia, Loeselia, Polemonium, Phlox, Gilia* and *Ipomopsis*. No one of these genera preponderates. Their aggregate total is 27 species, of which 41% are endemic. Within this area *Cobaea* reaches its northern limit of distribution and *Polemonium, Phlox, Gilia* and *Ipomopsis* reach their southern limit on the North American continent.

E. *Northern Sierra Madre.* Three tribes, the Bonplandia, Polemonium and Gilia tribes, are represented here by a total of ten genera and 39 species. *Ipomopsis* and *Phlox* lead in numbers of species. *Loeselia* reaches its northern limit; *Allophyllum* and *Eriastrum* reach their southernmost limits; and *Microsteris, Linanthus* and *Gilia* sect. *Arachnion* their southern continental limits within this area. A small percentage of the species (10%) are endemic

F. *Colorado Plateau.* In this area a large number of species, distributed among the genera *Polemonium, Collomia, Gymnosteris, Phlox, Gilia, Ipomopsis, Navarretia, Leptodactylon* and *Linanthus*, find their southern distributional limits. The area includes the northern and eastern limits of several *Linanthus* species. The total number of species is 60, composed very largely of *Phlox, Ipomopsis* and *Gilia*. Endemism attaining a frequency of 17% is found in these three genera.

G. *Sonoran Desert.* A single tribe, the Gilieae, is represented by five genera and 16 species. *Langloisia* reaches its southern limits within the area. Two species, or 12% of the rather scanty flora of 16 species, are endemic.

H. *Mojave Desert.* The genera *Gilia, Linanthus* and *Eriastrum* preponderate in numbers of species. Several otherwise cismontane Californian species of *Gilia, Linanthus, Eriastrum* and *Navarretia* reach their easternmost limits in the Mojave Desert. The flora of 41 species is quite rich for an area of relatively small size. Endemism, concentrated predominantly in *Gilia*, reaches 27%.

I. *Great Basin.* This is the only area in which all genera of both the Polemonium and Gilia tribes are present. The two tribes are approximately equal in numbers of species represented. The leading genera in order of species density are *Phlox, Gilia, Ipomopsis* and *Polemonium*. The total number of species is 73 but only 9 of these, or 12%, are endemic. The Great Basin is thus not a highly endemic area, but is, on the contrary, a meeting ground of Rocky Mountain, Southwestern and Pacific coastal Polemoniaceae.

J. *Cismontane Southern California.* This area comprises the coastal plains and mountains of southern California and northern Baja California. Here 54

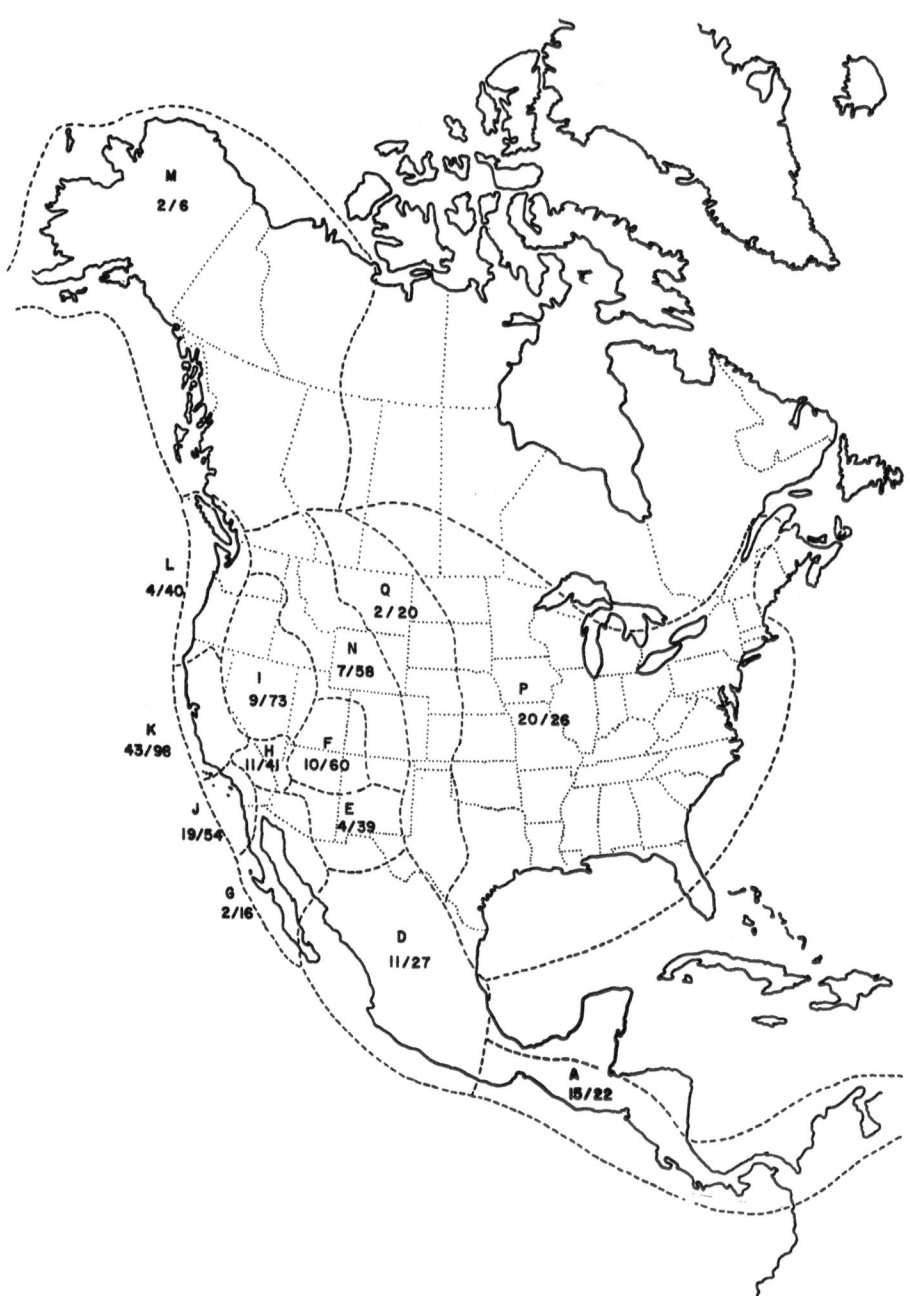

Fig. 69. Natural areas of the Polemoniaceae in North America.
The fractions indicate the number of endemic species and the total number of species in each area. For identification of the areas see text.

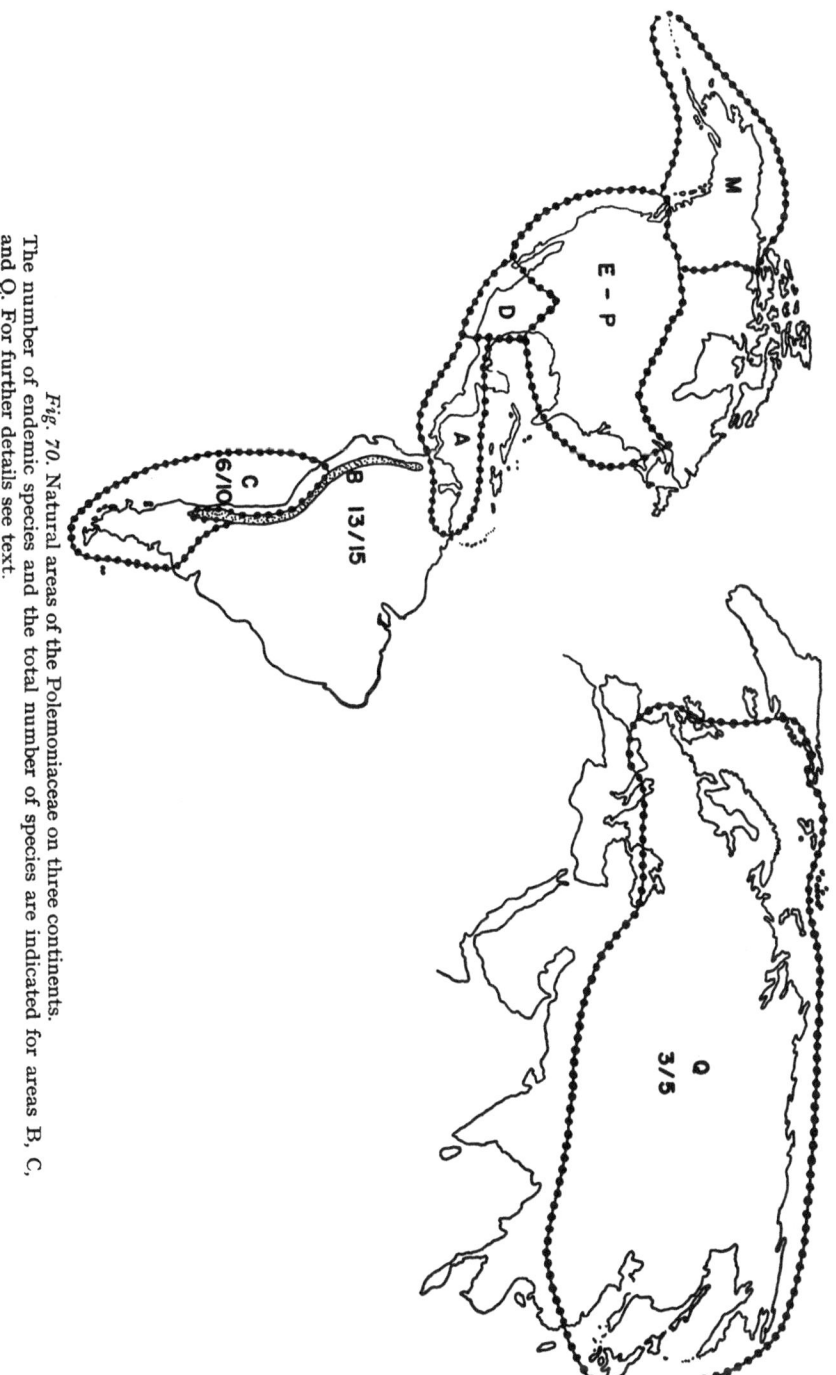

Fig. 70. Natural areas of the Polemoniaceae on three continents. The number of endemic species and the total number of species are indicated for areas B, C, and Q. For further details see text.

species distributed among every temperate genus except *Gymnosteris* are found. *Gilia* and *Linanthus* are easily the predominant genera both in numbers of species and in numbers of endemics. The percentage of endemism at the species level is 35%.

K. *Northern California*. The Pacific drainage region of central and northern California has the richest polemoniaceous flora of any area. There âre 98 species here. The leading genera in numbers of species are *Navarretia* and *Linanthus*, followed in second place by *Gilia* and *Eriastrum*. One genus, *Eriastrum*, finds its northern limits within this area. The number of endemic species, 43, is over twice as great as that in any other area, while the proportion of endemics, 44%, is also relatively high.

L. *Pacific Northwest*. A flora of 40 species, in which Collomias, Polemoniums and Phloxes are abundant, but various members of the Gilia tribe are also represented, is found in this area. The area represents the northern limit for the Gilia tribe and for *Allophyllum, Collomia* and *Microsteris* in the Polemonium tribe. The percentage of endemism is low, 10%.

M. *Alaska and Northwestern Canada*. Only six species, and these concentrated entirely in the genera *Polemonium* and *Phlox*, occur in this vast area. Four of these species find their continental limits of distribution within the area while two species of *Phlox*, or 33% of the total flora, are endemic here.

N. *Rocky Mountains*. Fifty-eight species about equally divided among the Polemonium and Gilia tribes make up the polemoniaceous flora of this region. *Phlox* is by far the most richly developed genus, though *Polemonium* and *Gilia* are well represented. The genera *Gymnosteris* and *Microsteris* and many species in all the other genera find their eastern limits in this area. Seven endemic species in *Phlox, Polemonium* and *Gilia* account for 12% of endemism.

O. *Great Plains*. The area lying east of the Cordillera and stretching from Canada to Mexico contains 20 species distributed about equally between the Polemonium and Gilia tribes. The leading genera are *Phlox* and *Ipomopsis*. Only two species of *Phlox* are peculiar to the area. The genera *Navarretia, Leptodactylon* and *Linanthus* reach their easternmost limits here. There is a marked floristic difference between the northern and southern parts of the Great Plains but no place to draw a boundary line between the two halves. In general the Great Plains does not possess a distinctive polemoniaceous flora of its own but contains instead an assemblage of outlying species derived from the adjacent floras of the Rocky Mountains and the Southwest.

P. *Eastern North America*. The large area extending through the prairie and forest regions of North America to the Atlantic coast is inhabited by some 26 species of Polemoniaceae. The leading genus, *Phlox*, contributes 2] of these species, while *Polemonium* is represented by two species and *Collomia, Gilia* and *Ipomopsis* each by one species. The flora is very different from that of the adjacent Great Plains, with only one species of *Collomia* and a few species of *Phlox* common to both regions. The endemism of 77%, concentrated chiefly in *Phlox*, is one of the highest in any area.

Q. *Eurasia*. A sparse polemoniaceous flora of five species belonging to two genera, *Polemonium* and *Phlox*, characterizes the Eurasian continent. Three of these species, or 60% of the total, are endemic.

It is clear from the above account as well as from the maps (Figs. 69, 70) that no direct correlation exists between the size of a natural area and the richness of its polemoniaceous flora. Two of the largest areas, Eurasia and the Alaskan region, contain five and six species respectively, whereas the two smallest regions, the Mojave Desert and Cismontane Southern California, have 41 and 54 species. Southern California has twice as many species and northern California about four times as many species as the entire eastern United States. The eight natural areas in the western United States all possess 38 or more species; the remaining areas throughout the world have 27 or less species. The species density is associated with the geographical position rather than with the size of an area.

The only natural area having one or more endemic genera is the Andean region. In terms of numbers of endemic species present, the main centers for the Phlox family are Northern California, Cismontane Southern California and Eastern North America (with 43, 19 and 20 endemics respectively). The chief endemic centers as measured by the percentage frequency of endemic species are the Andes (87%), Eastern North America (77%), Central America (68%), Temperate South American Lowlands (60%) and Eurasia (60%). It will be noted that these areas are all peripheral in relation to the main centers o species density. A moderately high frequency of endemics in combination with a rich floristic development is found in Northern California (43/98 = 44%), Cismontane Southern California (19/54 = 35%), and the Mojave Desert (11/41 = 27%). These three form the outstanding endemic areas within the region of high species density.

No species, genus or tribe occurs throughout the whole distribution area of the Polemoniaceae. The most cosmopolitan species in the family is *Polemonium caeruleum*, which occurs in eastern North America, western North America, the Alaskan region, and through Eurasia to Great Britain. *Microsteris gracilis* is found in seven natural areas in North America and one additional area in South America. *Collomia linearis* is represented in eight natural areas of North America. *Loeselia glandulosa* ranges from northern Arizona through Mexico and Central America to Venezuela.

At the opposite extreme, 181 species or well over half the members of the family are endemic to one or another natural area. All of the species of

Cobaea, Cantua and *Huthia* are confined to a single natural area. Of the species constituting the genera *Navarretia, Eriastrum* and *Phlox*, 63 to 66% are endemic within a single area. For *Bonplandia, Polemonium, Collomia, Gymnosteris, Gilia* and *Linanthus* between 50 and 57% of the species are endemics; for *Langloisia* and *Ipomopsis*, 40 or 42%; for *Leptodactylon* and *Loeselia*, 33 and 22%. Only the minor genera *Microsteris* and *Allophyllum* have no endemic species. Two genera (*Cantua* and *Huthia*) and two sections (in *Cobaea* and *Loeselia*) are endemic as a whole to single areas.

This means that the composition of the polemoniaceous flora changes in passing from one natural area to another. The phytogeographic picture must be painted in more than one or two dimensions. We must know, not only the number of species and the frequency of endemics, but also the kinds of species, for each natural area. We must compare the relative importance of the different taxonomic groups in the various areas.

In order to make this comparison as simple as possible, the tribes were taken as the units of measurement. For a given natural area the number of species belonging to each tribe (Table 12, center columns), was expressed as a percentage of the total number of species of Polemoniaceae present in that area. The percentages were then converted to sectors of a circle and represented graphically as a so-called "pie diagram." The proportionate representation of the different tribes in the 17 natural areas is shown on a map in Figure 71.

Certain floristic relationships are brought out by the map. In the first place, the five tribes of the Polemoniaceae do not meet in any one spot, though they all occur on the South American continent. Four tribes occur together in the Mexican Highland, and three tribes coexist in two areas, the Andes and the Northern Sierra Madre. In the northern and southern temperate zones only two tribes, the Polemonieae and Gilieae, are found.

The Cobaea tribe preponderates in Central America, diminishing in importance in the Mexican and Andean regions. The Bonplandia tribe is well developed in both Central America and the Mexican Highland and declines gradually to the north and south of its center. The Cantua tribe is confined to the Andes.

The Gilia tribe preponderates in the warm arid regions of western North America. It is the only tribe present in the Sonoran Desert and almost the only tribe in the Mojave Desert. North-south transects in North America, whether on the Pacific slope or in the intermountain region, show a regular decline in frequency of this tribe from lower to higher latitudes. The Gilia tribe is only sparingly represented in eastern North America, where it is confined to the southern half of the area, and is absent entirely from northwestern Canada, Alaska and Eurasia.

The Polemonium tribe, on the other hand, preponderates in the colder and

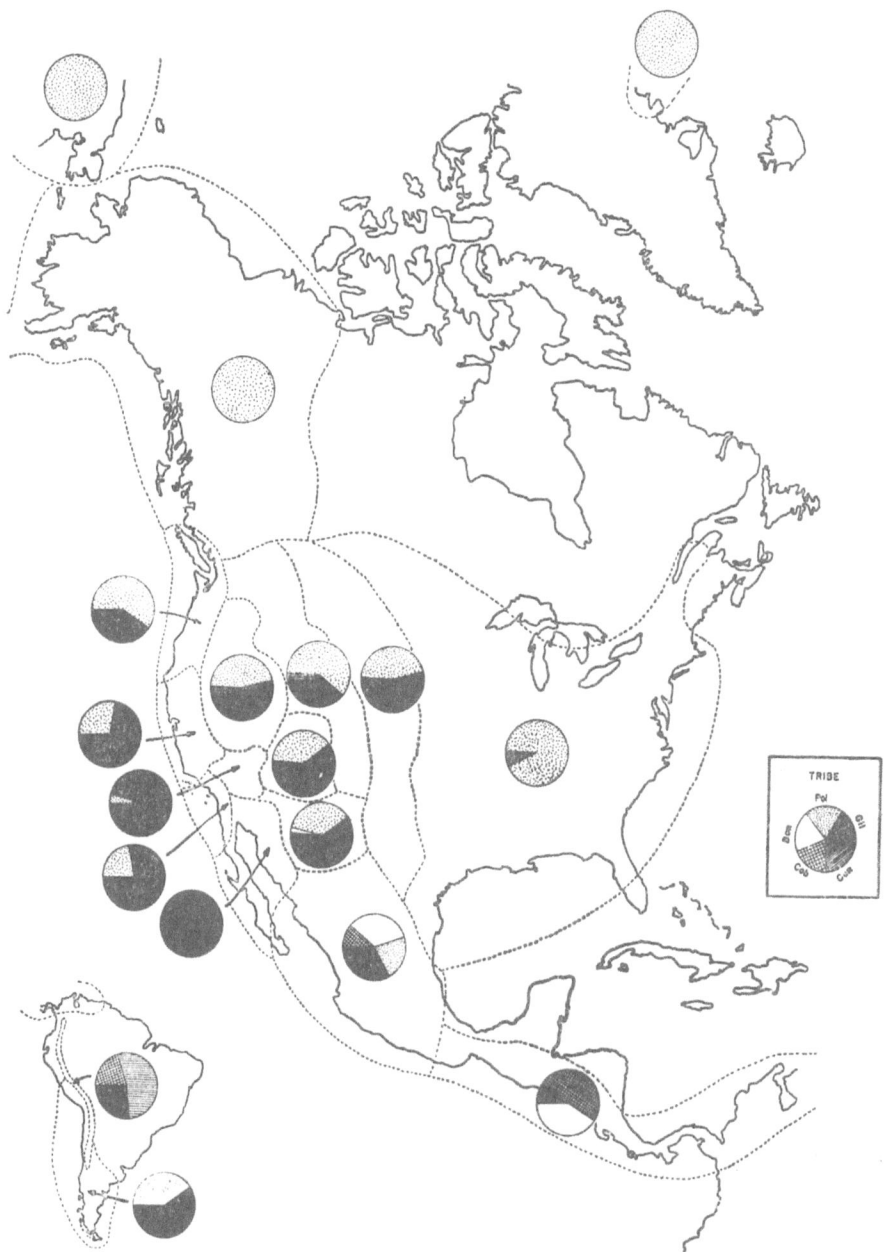

Fig. 71. Composition of the polemoniaceous flora in the different natural areas.
For each area the proportion of the species belonging to each tribe is indicated by a sector in the
pie diagram. (Pol = Polemonium Tribe; Gil = Gilia Tribe; Can = Cantua Tribe; Cob =
Cobaea Tribe; Bon = Bonplandia Tribe.)

moister regions. It is the only tribe represented in the Alaskan and Eurasian provinces, the dominant tribe in eastern North America and the major tribe in the Pacific Northwest and Rocky Mountains. It becomes a relatively minor element in the polemoniaceous floras of the arid southwestern deserts. Within the Great Plains area, which taken as a whole has an approximately equal distribution of the Polemonium and Gilia tribes, the former preponderates in the northern and the latter in the southern parts.

For the Polemonium tribe the greatest concentration of species lies in the Rocky Mountains and Great Basin, for the Gilia tribe in the California region. The two temperate tribes, therefore, like the tropical tribes Cobaeeae and Cantuaeae, have different centers of distribution.

CENTERS OF ORIGIN

The present distribution areas of the Polemoniaceae are the product of a long period of evolution and migration. The facts set forth in the preceding section raise the question of how the actual geographical patterns arose. Where did the major phylads of the Phlox family originate? By what steps did the family come to occupy its present centers of distribution?

Direct evidence bearing on these questions is lacking. There is just one fossil record of the Polemoniaceae[1]. This is a fruit of a recent species, *Phlox sibirica* or *P. borealis* or a close relative, from a deposit of Pleistocene age near Fairbanks, Alaska (Chaney and Mason 1936). What it tells us is that a *Phlox* species which is present in that area today was also there some tens of thousands of years ago. This information is clearly of little relevance for the broader problems of migrational history in the Phlox family.

Nevertheless, the task of determining the approximate centers of origin and lanes of migration is not entirely hopeless. Several lines of indirect evidence are available. We can identify the present broad floristic associations of the existing Polemoniaceae and, assuming that these associations have prevailed for a long time, proceed to trace out the known geological history of the parental floras. The floristic history of the areas where the Polemoniaceae are now well developed has been summarized by Axelrod in his important review paper of 1950, on which the following account is based. We can also correlate the probable phylogenetic trends within the family, as set forth in Chapter 7, with the climatological and geological history of the various natural areas occupied by the family.

The conclusions arrived at by this indirect method of inquiry will obviously not have the certainty of a chromosome number determination, of a distribution map, or even of a postulated history of floras based on fossil records.

[1] See footnote on page 194.

If, however, the lines of evidence available to us are consistent among themselves we may accept the resulting picture of the development of the Phlox family in space and time with a fair degree of confidence. In science we should go as far as we can with the facts at hand; in the present instance this limit of knowledge is represented by a probable but unverifiable hypothesis.

For the most part the evidence does not justify attempts to identify a center of origin in precise geographical terms. The finding of a primitive species in a given area does not mean that the group derived from this species originated in this area, as Matthew (1939) so clearly pointed out. The geographical area of a species may shift through time with changing climatic conditions, while the broad ecological tolerances remain relatively constant. Migration is usually easier than evolution. And the environment migrates as well as the species.

It follows that we may be able to establish centers of origin in a broad floristic province but unable to locate that center in any particular geographical area. On the basis of present distributions, Wherry has postulated centers of origin in north-central Canada for *Polemonium* sect. *Polemoniastrum, Collomia* and *Phlox* (Wherry 1940, 1944a, 1944b, 1955). Eschewing such uncertain deductions, I have been content to conclude that these genera arose somewhere in the Arcto-Tertiary forest.

The Phlox family is today a predominantly New World family and it is reasonable to seek its center of origin somewhere in the region of its present abundance. In the absence of fossil records, however, we can never be certain that the family did not originate in some other part of the world. For all we know, the earliest Polemoniaceae diverged from the ancestral tubiflorous stock in Australia or India or Africa and later disappeared without leaving any trace in the flora of its original home. No one would guess from recent floristic affinities alone that the Breadfruit-tree ever existed in the California flora, or that *Ephedra* was present in northern Europe as recently as the Late Pleistocene, or that the Proteaceae once occurred widely throughout North America and Eurasia.

Even if the Phlox family did originate outside the American hemisphere its major development, which it is our purpose to explain, took place in a New World setting. The task before us is to reconstruct from known geographical relationships the probable historical changes which have occurred since the origin of the American Polemoniaceae. The events which occurred prior to the rise of the American genera are unknown, perhaps unknowable, and in any case irrelevant for the task at hand.

It was concluded in Chapter 7 that the most primitive characters for the Phlox family are found in the tropical genera *Cobaea, Cantua, Huthia,* and *Bonplandia. Cantua* and *Cobaea* in particular possess many primitive morphological features, whereas the other genera named above along with *Loeselia* and *Polemonium* possess characters that might have developed by divergence from an ancestor like *Cantua* in some respects and like *Cobaea* in others.

Cobaea is today restricted to montane tropical rainforest where a mild equable climate prevails, and some species of *Cantua* inhabit tropical woodland. The environments corresponding to these vegetation types have existed continuously during the entire period of angiosperm evolution and have usually been more widespread than they are today. Other environments occupied by Polemoniaceae – the hot deserts, frozen tundras, high mountains – have had their greatest expansion in recent times. Insofar as they inhabit an ancient type of environment, *Cobaea* and *Cantua* probably exhibit a primitive ecology paralleling their primitive morphology.

We will assume on this basis that the common ancestor of *Cantua* and *Cobaea* was present in the lower Tertiary, viz. during the Eocene or Early Oligocene period, in some facies of the old tropical forest which then covered a wide area from South America to southern North America.

During the Late Oligocene and Early Miocene the northern arm of this neotropical flora developed under the influence of more arid climatic conditions into a xeric type of vegetation. The woodland vegetation composed of subtropical trees, shrubs and grasses adapted to a warm semi-humid semi-arid climate is known as the Madro-Tertiary flora (cf. Axelrod 1950a, 1950b). By Middle Miocene time it spread as far north as southern California.

The Bonplandia tribe, which is phylogenetically a derivative of some tropical Polemoniaceae having many of the characteristics of the modern Cobaeas, occurs today in the subtropical forests and xeric woodlands of Mexico and Central America. It is probable that the divergence of this tribe from the ancestral Polemoniaceae involved a shift in ecological zone from the neotropical rainforest to the more xeric Madro-Tertiary flora, i.e., quantum evolution in the sense of Simpson.

Similar changes to a drier climate took place during the Oligocene and Miocene on the southern margin of the neotropical flora in South America. Here as in North America the result was a replacement of the tropical forest by more xeric subtropical woodland and scrub. Some members of the Cantua tribe are associated today with an open xerophytic formation of columnar cacti and rain-green shrubs in the Andes. The divergence of the xeric Cantueae from their more primitive rainforest-inhabiting ancestor, like the parallel divergence of the Bonplandieae in the northern hemisphere, probably required an ecological shift from a tropical to a subtropical environment.

Through Miocene and Pliocene time the climate continued to become increasingly arid. Keeping pace with these changes, the more humid phase of the Madro-Tertiary flora gradually gave way to a more arid phase in western North America. Many evergreen oaks, thorn forest species, and sub-tropical trees and shrubs, which had been represented in western North America during the Miocene, retreated to Mexico by Middle Pliocene time, while other elements of the Madro-Tertiary flora, such as pinyon pine, juniper, live-oak, chaparral species, and plains grasses, remained in possession of the western American territory. In the Middle Pliocene the northern branch of the Madro-Tertiary flora broke up into segregates: pinyon-juniper woodland in the areas with cold winters; live-oak woodland in areas with mild but moist winters and dry summers; short-grass plains in the interior areas with low rainfall; and chaparral in warm dry zones.

The climatic and floristic associations of the earliest Gilieae are plainly suggested by their tendency at the present time to reach their best development in warm arid regions. When we examine the vegetation types with which the Gilia tribe is associated we find that nearly every North American species in the tribe occurs in one or more of the present northern segregates and remnants of the old Madro-Tertiary flora: the pinyon-juniper woodland, live-oak woodland, chaparral, and short-grass plains. The Gilieae are found in other vegetation types too, in the deserts, alpine zones, and coniferous forests, but the association with the Madro-Tertiary flora is like a common denominator running through the whole tribe. This association involves the more primitive groups in each genus and section, whereas the associations with other vegetation types can be viewed as later invasions by more specialized species and races in the different lineages.

As noted in Chapter 7, no single existing member of the Gilia tribe exemplifies the ancestral condition of the tribe, though primitive characters are found within the genera *Gilia*, *Ipomopsis*, *Eriastrum* and *Leptodactylon*. By pooling the primitive features still preserved and extrapolating from them we concluded that the ancestor of the Gilia tribe was a small alternate-leaved xeric shrub related to the modern Loeselias. The most probable inference from the considerations presented above is that this unknown ancestor of the Gilia tribe diverged from a *Loeselia*-like stock within the Madro-Tertiary flora. This divergence represented another shift in ecological zone, in this case from the subtropical to the arid phase of the Madro-Tertiary flora.

In the segregation of the Madro-Tertiary flora into its present-day communities, which occurred under the influence of increasing aridity and a climatic differentiation in the Middle Pliocene, different phylads within the Gilia tribe retained associations with different segregate communities. The genera *Eriastrum*, *Navarretia*, *Linanthus*, and the section *Gilia* of *Gilia* remained with the California live-oak woodland and digger-pine savanna. *Ipomopsis* and *Gilia* sect. *Giliandra* associated themselves chiefly with the pinyon-juniper woodland, and *Gilia* sect. *Giliastrum* with the short-grass plains and xeric subtropical woodland. The expansion of arid habitats now permitted these groups to undergo a rich speciation and development in their respective regions.

As we survey the present floristic associations of the Polemonium tribe we cannot help but note that this tribe is only a casual member of the present remnants of the Madro-Tertiary flora. Its basic associations seem to be with the various modern derivatives of the temperate Arcto-Tertiary flora which covered the northern half of North American in Tertiary time. Thus *Polemonium* and *Phlox* both occur in the eastern deciduous forest, the western coniferous forest, the temperate forest of Mexico, and the arctic tundra. *Collomia* is present in the Pacific coniferous forest and *Gymnosteris* in the sagebrush plains. These are all segregates of the old Arcto-Tertiary flora. The associations of the Polemonieae with Madro-Tertiary floras can be regarded as later invasions by advanced members belonging to several of the genera.

The Arcto-Tertiary forest was in contact with the neotropical forest during the Eocene and has been in contact with the Madro-Tertiary flora since the late Oligocene. The Polemonium tribe is believed to be derived from a *Cobaea*-like or *Bonplandia*-like ancestor existing in the neotropical forest or the humid phase of the Madro-Tertiary forest. Its divergence from the tropical or subtropical ancestor must, consequently, have entailed an ecological shift from tropical or subtropical to temperate conditions.

We find the Polemonium tribe growing today in eastern North America, on the highest western mountain peaks, in Eurasia, and in the tundra around the pole. Obviously the ecological shift which accompanied the rise of the Polemonium tribe was not a sudden transition from a tropical rainforest to an arctic tundra environment. It is more likely that the evolutionary transition from one ecological zone to another occurred at a point where the gap was narrowest. A shift from a cloud-forest environment to a warm temperate environment seems feasible. In this connection it is worth noting that *Cobaea*, representing a tropical form, today occurs within the territory of *Polemonium*, which represents a primitive type of Polemonieae. The montane rainforest zone of *Cobaea* and the temperate forest zone of *Polemonium* occur in juxtaposition in the Mexican highland. A shift from one zone to the other under conditions such as these would suffice to start the derivative group on a new course of evolution, leading eventually to the occupation of more extreme environments.

The idea that large parts of the western North American flora have had a Mexican origin was expressed by several floristic botanists of the late 19th century, by Hooker, Hemsley, and Gray. In the present century the same view was advocated by Harshberger, Clements and others[2].

This postulation was at first based on floristic relationships. Thus Hemsley showed that the flora of southern Mexico is quite different from that of northern Mexico. The former is related to some extent to the eastern North American flora. With regard to the latter, he states: ". . . North Mexico is the centre of a special xerophilous flora, which, there are good grounds for assuming, originated in this area, though this flora now has considerable northward and southward extensions." This Mexican plateau flora, according to Hemsley, extends into western North America, so that of 560 genera in Arizona 402 are common to Mexico, and of 1473 species in Arizona 429 are common to Mexico.

Gray and Hooker (1880) stated flatly that: "A large part of the botany of California, still more of Nevada, Utah, and Western Texas, and, yet more, that of Arizona and New Mexico, may be regarded as a northward extension of the botany of the Mexican plateau." In a later paper Gray (1884) wrote: "It is now becoming obvious that the Mexican-plateau vegetation is the proximate source of most of the peculiar elements of the California flora, as also of the southern Rocky Mountain-region and of the Great Basin between; and that these plants from the south have competed with those from the north on the eastward plains and prairies."

There were some figures to support this floristic relationship in the time of Gray, Hooker and Hemsley. Even better statistics are available today. Kearney and Peebles (1951) have shown that 36% of the higher plant species in the Arizona flora belong to the Mexican flora.

The elements in the western North American flora believed to be derived ultimately from a Mexican flora include the cacti, agaves, yuccas, *Fouquieria*, *Fremontia*, *Lupinus*, Onagraceae, Nyctaginaceae, Boraginaceae, Hydrophyllaceae, some xerophytic grasses such as *Bouteloua* and *Sporobolus*, many xerophytic Compositae, and not least, the Polemoniaceae.

The taxonomic evidence for a floristic connection between western North America and northern Mexico was supplemented in 1939 by paleobotanical evidence. Axelrod (1939) analyzed a Middle Miocene flora from near Tehachapi in southcentral California. He showed that this fossil flora consisted of elements some of which are now scattered over the Southwest but the great majority of which are now found in northern Mexico. The fossil evidence indicated a spreading through the Southwest during the Miocene of a wood-

---

[2] Gray and Hooker 1880; Gray 1884; Hemsley 1886–1888: 309; Harshberger 1911: 182, 285, 294, 334; Abrams 1915, 1925; Clements 1920: 145, 165, 179, 196; Jepson 1923: 3; Munz 1935: xv; Axelrod 1938, 1939, 1940, 1950b; Kearney and Peebles 1942: 7, 1951: 23; Mason 1942; Campbell and Wiggins 1947.

land vegetation adapted to a hot and semi-arid climate (Axelrod 1938, 1940). This assemblage was the Madro-Tertiary flora.

It is generally agreed, therefore, that a close connection, both contemporaneous and historical, exists between the floras of the Mexican plateau and western North America. The interpretation usually placed on these facts, as we have just seen, is that a large part of the western flora originated in the Mexican highland area and migrated from there to the present distribution areas. The facts justify only the conclusion that the Madro-Tertiary floristic elements originated in one or more areas possessing climatic conditions like those now prevailing in northern Mexico. The geographical center (or centers) of origin may or may not have been in Mexico. On the floristic and paleontological evidence available we can identify with a reasonable degree of reliability a broad climatic, but not a specific geographical, center of origin of the western flora.

As regards the Polemoniaceae, the hypothesis proposed here is that an ancestral form, existing perhaps in the Mexican highland, or perhaps somewhere else, was adapted to a tropical climate characterized by a relative abundance of both moisture and warmth. This ancestral form gave rise to two lines which were to have a conspicuous evolutionary success in North America. The two lines came into being as adaptations to new climatic conditions which developed following the breakup of the pre-existing mild climates over extensive areas. The conditions which became widespread during the Tertiary were an increase in aridity and an increase in cold.

The Gilia tribe retained the ancestral tolerance for warmth but was forced to sacrifice the ancestral requirements for abundant moisture in order to fit into the warm arid climates then developing. The Polemonium tribe, on the other hand, retained the ancestral requirements for moisture, while giving up the warm temperature tolerances as a condition for fitting into the new cold humid climates. The Gilia tribe underwent its major development as a component of the Madro-Tertiary flora. The Polemonium tribe was captured from its tropical or subtropical ancestral home at an early stage by the Arcto-Tertiary flora.

The two tribes overlap broadly in distribution today. Members of the Gilia tribe occur in coniferous forest and members of the Polemonium tribe in xeric woodland. The Gilieae and Polemonieae grow side by side in numerous places throughout the west. This overlap is a part of the general intermingling of the two great North American floras, the Arcto-Tertiary and Madro-Tertiary, which over the ages has led to an exchange of many floristic elements.

ORIGIN OF THE DESERT ANNUALS

The desert area of North America at the present time falls into four climatically and floristically distinct regions. As outlined by Shreve (1942;

cf. also Fig. 69), these regions are: the Great Basin, which includes most of Nevada, western Utah, eastern Oregon, and southern Idaho; the Mojave Desert, stretching from southeastern California to the southwestern corner of Utah; the Sonoran Desert, composed of the adjacent parts of Sonora, Baja California, extreme southeastern California, and Arizona; and the Chihuahuan Desert, of Trans-Pecos Texas, Coahuila, and Chihuahua.

These deserts occupy areas which during the Tertiary supported grassland, woodland, and even forest (Axelrod 1950b). A climatic trend toward lower rainfall accompanied by the rise of the Cascade-Sierra Nevada-San Bernardino mountain axis along the west coast culminated in the origin of the interior desert during Pliocene time. The preexisting woodland and grassland formations were replaced by desert in the Late Pliocene. The Great Basin Desert seems to have differentiated out of the Arcto-Tertiary Flora, while the warmer southern desert emerged from the Madro-Tertiary Flora. The area of the warm southern deserts was floristically uniform during the Pliocene and is believed to have become differentiated into the present Mohave, Sonoran and Chihuahuan provinces during Pleistocene time.

The appearance of the desert habitats made possible a new wave of adaptive evolution. It is of interest to examine how the Phlox family responded to the challenge of the new desert environments.

The Polemoniaceae include some of the most abundant and characteristic members of the spring flora in the present desert areas. This family is particularly abundant in the Mojave Desert but is well represented in the Great Basin and Sonoran Deserts. With but few exceptions, and these confined almost entirely to the genus Phlox in the Great Basin, the desert Polemoniaceae are annual herbs. Some 20 taxonomic groups–sections in the case of large genera and whole genera in the case of small genera–contribute to the polemoniaceous desert flora. Of these 20 taxa, six are members of the Polemonium tribe and fourteen belong to the Gilia tribe.

It is significant that only one of the 20 groups, Langloisia, is endemic to the desert. Two other groups (*Gymnosteris* and *Linanthus* sect. *Dianthoides* are very nearly endemic in the desert. The remaining taxa are represented not only in the deserts but also in the geologically older plant associations bordering the deserts, such as pinyon-juniper woodland, chaparral, and xeric yellow pine forest. In a number of cases a single species embraces both desert and non-desert races. *Microsteris gracilis*, for example, inhabits both the yellow pine forest and the sagebrush plains of the Great Basin Desert. *Gilia ochroleuca* occurs in pinyon-juniper woodland and chaparral as well as in the Mojave Desert.

These facts suggest strongly that as the desert habitats opened up they were occupied by different phyletic lines already existing in the bordering

woodland and forest communities. The occupation could have been accomplished either by an invasion from adjacent areas or by a progressive
evolutionary transformation of the populations in situ.

The desert races and species often show reduced characteristics when
compared with their closest non-desert relatives. The desert race of *Gilia
ochroleuca* is small-flowered and self-pollinating, whereas the woodland races
of this species are large-flowered and outcrossing. The desert forms of *Microsteris gracilis* are more strongly reduced than the pine forest types. Several
desert species of *Linanthus* and *Eriastrum* possess vegetative and floral
characters which are clearly derived by reduction from characters similar
to those found in related species of bordering communities. Facts such as
these, which could be multiplied indefinitely, furnish additional support for
the thesis that the desert forms have been derived phylogenetically from
ancestors adapted to conditions outside the deserts.

The majority of the desert annuals belong to the Gilia tribe. The members
of this tribe are today concentrated chiefly in the California region where they
are very abundant both in numbers of species and numbers of individuals.
They diminish markedly in frequency in the bordering areas. What factors
determine these limits of distribution?

One critical factor is the nature of the substrate. The Gilieae as a whole
require coarse well-drained soils and thrive best on slightly compacted sand.
It is on such substrates that the Gilieae are usually found in California and in
neighboring areas. (There are of course a few exceptions, such as the vernal
pool Navarretias which grow on mud or dried clay.) Sandy areas suitable
for the majority of the Gilieae are very common in California, abundant in the
southwest, and relatively uncommon over large parts of the Chihuahuan
Desert, Colorado Plateau, Great Basin, Rocky Mountains, Great Plains and
Pacific Northwest.

In California the populations of *Gilia, Eriastrum, Linanthus* and other
sand-inhabiting Gilieae are crowded close together. This is true of both the
desert and the non-desert forms. In Oregon, Wyoming, Utah and several
other western states the populations of a species exist in a much more diffuse
condition, neighboring colonies being separated by scores or even hundreds
of miles of unfavorable terrain. The only sandy habitats in extensive tracts
of plains may occur on the banks of major rivers or in occasional sand hills.
Here and here alone are the Gilieae of California affinity found. The density
of the Gilia tribe in California and its more scattered distribution in other
sections of western North America are reflections in part of the dense or
scattered occurrence of suitable sandy habitats in the different areas.

Sandy soils do not constitute a limiting factor throughout large areas of the
southwest. Yet a number of annual species of the Gilia tribe extend east

from California through the southwestern deserts only to drop out beyond the El Paso region of Texas. The Cobwebby Gilias (*Gilia* sect. *Arachnion*), for example, a predominantly desert phylad, are prominent in the Mojave Desert, sparingly represented in the Sonoran Desert, occasional in the El Paso region, and altogether absent from the Davis Mountains region of western Texas and from the Chihuahuan Desert of Mexico. The genus *Eriastrum* and the sections *Dactylophyllum* and *Linanthus* of *Linanthus* follow a similar pattern. What factors come into play here?

A critical factor in these cases appears to be the presence or absence of winter rains. The annual Gilias, Linanthuses and some other Gilieae have a life cycle geared for seed germination and early seedling development under the stimulus of winter rains followed by vegetative growth, flowering and fruiting under the stimulus of rising temperatures and falling soil moisture in the spring. The optimum conditions for seed germination in these plants have been shown by Went, Grant and others to include distinctly cool temperatures (cf. bibliography under the subheading of Ecology).

Winter rainfall and a warm spring are characteristic features of the California climate where these members of the Gilia tribe are most abundantly developed. Some winter rains fall on the sandy areas of the Sonoran Desert in southern Arizona and support a sparse flora of annual Gilieae. The sandy habitats of the El Paso region are drier still and the annual flora even sparser. Indeed the question arises whether the period of seed viability of most annual species of Polemoniaceae is sufficiently long to enable them to tide over the years of drought between favorable seasons in this area.

The available sandy habitats farther east in the Davis-Chisos Mountains area of Texas are not occupied by any representatives of the California phylads of annual Gilieae. The winters in this area are cold, long and dry. The slight winter precipitation, coming in the form of snow, has largely disappeared from the upper levels of the soil by the time the soil and air temperatures have reached the threshhold for plant growth. This sequence of temperature and moisture conditions is a far cry from the climatic pattern found in California, and it lies wholly outside the tolerance range of the annual Polemoniaceae derived from the California flora.

Most of the desert annuals belonging to the Gilia tribe have their closest systematic relationships with the Gilieae of cismontane California. They have ecological requirements–preferences for sandy soils, cool winters, warm growing seasons, and winter rainfall–corresponding to conditions that are widespread and typical in the California foothills and valleys. These requirements were probably built into the ancestors of most of the desert annual Gilieae by adaptation to California climatic and edaphic conditions. The floristic affinities and the ecological requirements of a majority of the desert

Gilieae both point to a center of origin in an area which possessed an environment like that of cismontane California today. It is altogether possible indeed that this original homeland of many of the desert Polemoniaceae was actually California.

These elements are today richly developed in the Mojave Desert, sparingly represented in the Sonoran and Great Basin Deserts, and virtually absent from the Chihuahuan Desert. The numbers of individuals and the numbers of species show a progressive decline in frequency as the environmental factors become increasingly different from the cismontane California conditions.

A likely interpretation of these facts is that certain broad adaptations built up in the ancestral stocks in relation to a California environmental complex–an annual life cycle with winter seed germination, growth in sandy substrates, and a spring season of flowering–predisposed or "preadapted" these stocks for a successful transition to the similar environmental conditions which developed in the Mojave Desert. These same predispositions enabled a few of the species to colonize the less favorable habitats developing in the Great Basin and Sonoran Deserts. But the radically different conditions in the Chihuahuan Desert lay beyond the limits of their actual or potential tolerance range.

Not all of the desert Gilieae have California affinities. Several species of southwestern desert annuals belonging to *Ipomopsis* sect. *Microgilia* (*I. gunnisonii*, *pumila*, *sonorae*, etc.) have Great Basin and Rocky Mountain relationships. Desert species of *Gilia* sect. *Giliastrum* in Trans-Pecos Texas, the Southwest, and Mexico are apparently derivatives of a Texas woodland vegetation.

The desert taxa of the Polemonium tribe are with one exception, *Allophyllum*, confined to the cold northern Great Basin Desert. The conclusion seems justified that the desert Polemonieae, by contrast with the desert Gilieae, have evolved from ancestral stocks belonging to the Arcto-Tertiary flora.

PLEISTOCENE RELICTS

The climate at various times during the Pleistocene was colder and wetter than it had been previously or has been since. The ice descended four times from northern centers and covered the northern half of the North American continent (see Fig. 72). Pluvial periods developed in western North America, where continental glaciers never formed. The retreat of the ice following each glacial maximum was marked by warmer and drier interglacial stages.

The biogeographical and evolutionary effects of these changing conditions were undoubtedly very complex. The entire subject has been critically reviewed by Deevey (1949), who emphasizes the many lacunae in our knowledge of the Pleistocene in North America, the superficial and speculative

Fig. 72. The present distribution area of the genus Phlox in North America compared with the area covered by the latest (Wisconsin) ice sheet (stippled area). Note the relatively slight Post-Pleistocene migration of Phlox into the glaciated area. (redrawn with modifications from Wherry.

sharacter of many attempts to interpret modern distributions in terms of cupposed Pleistocene effects, and the potential richness of Pleistocene biogeography as a field for serious ad hoc research.

The phytogeographical consequences of the complex events of Pleistocene time are numerous and varied. A widescale extinction of populations in areas occupied by ice sheets must have taken place. The warm dry interglacial periods must have also taken their toll, as Wherry (1955: 6) has pointed out. The shifting of floristic zones with the changing climatic conditions must have frequently led to contacts between previously isolated species, while the withdrawal of ice, snow and water opened up habitats for the establishment of their hybrid derivatives. We have already seen that the differentiation of the deserts into the modern Mojave, Sonoran and Chihuahuan phases probably took place in the Pleistocene, and we shall see later that some amphitropical distribution patterns probably also originated during this period.

One effect which has been emphasized repeatedly since the time of Darwin is the development of arctic-alpine and boreal-alpine distributions and various other major and minor disjunctions of area. The expansion of cool and moist habitats permitted range extensions of many plant and animal populations to the southward across lowland areas. The retreat of the ice in a subsequent warm dry interglacial period was accompanied by a contraction of the distribution range. The plant or animal population did not only retreat to the north, however, but also withdrew to higher elevations in the nearby mountains, where it found similar climatic conditions. At the end of an Ice Age, a previously continuous distribution area might be broken up into discontinuous relictual areas isolated on separate mountain peaks.

Not all disjunct distributions, however, are necessarily relictual in origin. Isolated populations of a species may also arise as a result of long-range dispersal. Several alleged cases of Pleistocene relicts occupying areas not known to have been covered by ice or water during the Pleistocene probably belong to this latter category (cf. Deevey 1949). Nevertheless, when the disjunction is paralleled in many or most of the components of an entire community and includes species with poorly developed means of dispersal, the inference that it is relictual is reasonably well justified.

The following paragraphs are devoted to some interesting disjunct distributions in the Polemoniaceae which probably or possibly originated since the end of the Pleistocene. The cases to be mentioned are drawn almost exclusively from the western American flora, with which I possess some familiarity, though many equally interesting instances could be cited from the eastern United States and other parts of the world.

Some of the best examples of Pleistocene relicts in the Phlox family are to be found in the genus *Polemonium*. Davidson (1950), viewing the situation in the genus from the taxonomic standpoint, comments as follows. "The problem of delimiting species in *Polemonium* is complicated by their predominantly montane habitats. Mountain-top isolation of populations, comparable to insular isolation, favors a certain amount of random fixation of genes. In some such cases speciation may result in narrow endemics restricted to a single peak or mountain range. In other cases the populations on neighboring but isolated mountains are remarkably similar morphologically, a fact which presumably suggests their common origin. If cognizance were taken of the minute differences existing between populations of different mountains, one might eventually delimit as many subspecies as there are populations on isolated mountains."

Polemonium pulcherrimum ranges today from the Arctic Circle in Alaska through Alberta to the northern Rocky Mountains in Wyoming and through British Columbia and the Cascade Mountains to the central Sierra Nevada of California. In Alaska it is found at sea level, in British Columbia in the intermediate valleys, and in California in the alpine and subalpine zones of the mountains. Consequently *Polemonium pulcherrimum* exists in a series of isolated populations confined to different mountain peaks in the southern part of its range, whereas in the north "the genetic isolation is reduced to a minimum" (Davidson 1950). The insular populations in the southern mountains have probably arisen from a more continuous ancestral population present at lower elevations during a cooler climatic phase.

The western American subspecies of *Polemonium caeruleum*, known as *occidentale* or more properly as *amygdalinum*, has a main distribution area extending from the Yukon to Utah and the Sierra Nevada of California. A disjunct station of this subspecies exists far to the south in the San Bernardino Mountains. This population is isolated by an airline distance of about 200 miles from the closest Sierran population. The intervening territory is largely lowlands. In this case the entire assemblage of marsh plants and coniferous trees with which *P. caeruleum* is associated in its southern outpost has predominantly Sierran affinities. The most likely conclusion that can be drawn from these facts is that *P. caeruleum* and associated species were linked up with the main bodies of their species in the Pleistocene but separated since.

Polemonium caeruleum caeruleum of Eurasia has isolated peripheral populations in central England and the Pyrenees which are separated from the populations in the Alps and Scandinavia by distances of 450 or more miles. These gaps probably date back to Pleistocene times. That this plant was more widely distributed in Britain during the late Pleistocene is attested to by the fossil record, for its pollen has been found in late glacial sites scattered through southern England and Wales far outside its present restricted area

(Godwin 1956: 159). *Polemonium caeruleum vanbruntiae* apparently occurs in two disjunct areas of the Allegheny plateau in the eastern United States, a southern area in West Virginia and a northern area comprising Pennsylvania, New York and Vermont (Cain 1944: 462). Cain believes that this separation came about during a postglacial warm period.

Polemonium chartaceum has a remarkable distribution. This species occurs on high peaks in the White Mountains east of the Sierra Nevada and again on two mountain peaks in Siskiyou County, California, 320 miles distant. Whether these isolated colonies are remmants of a formerly more widespread distribution is not known.

Polemonium viscosum occurs through the Rocky Mountains from British Columbia to northern Arizona and New Mexico. A disjunct colony is found on high mountains in central Nevada. Farther west across another distributional gap the related *P. eximium* occurs on the high peaks of the Sierra Nevada. The distribution areas of these species were probably established during and since the Pleistocene.

Polemonium micranthum forms a widespread population system in the Great Basin as far south as Reno, Nevada. After a gap of 320 miles it recurs on Mt. Pinos in south-central California (Fig. 75a). It is associated with sagebrush in both its northern and southern areas. The disjunction in this case might be due to long-range dispersal, however, rather than to the dissection of a previously more continuous distribution area, since *P. micranthum* is a wide-ranging and apparently fairly mobile species. In South America a gap of about 600 miles exists between a northern population and the main area farther south. Future collecting may possibly disclose intermediate stations there.

The species pair, *Collomia mazama* and *C. rawsoniana*, exhibits a remarkable distribution pattern. The plants are perennial, broad-leaved and mesic, in which respects they are relatively primitive for the genus *Collomia*. In their distribution these primitive species are both narrowly endemic, although the habitats they occupy – moist stream-banks in mid-altitude coniferous forest – are widespread on the Pacific slope of North America. *Collomia mazama* is confined to the borders of several permanent creeks on the west slope of Mt. Mazama (Crater Lake) and again on Mt. McLoughlin. These two mountains lie some 30 miles apart in southern Oregon. The related *Collomia rawsoniana* is found only along a few streams in the south-central Sierra Nevada of California. Its total area is 15 miles long and several miles wide. A gap of 375 miles separates *Collomia mazama* and *C. rawsoniana*. The intervening territory contains numerous permanent streams in transition zone forest which, superficially at least, seem similar to the habitats occupied by the two species today.

A series of well marked morphological differences between *Collomia mazama* and *C. rawsoniana* rules out the possibility of a derivation of one disjunct population from the other by long-range dispersal in very recent times. The evolutionary conservatism of the plants suggests, on the contrary, that their differentiation from a common ancestor must have occurred in earlier times. In view of their mesic character, the age of the maximum development and spread of the group could have been some pluvial stage during the Pleistocene. The extreme endemism of each species today and the disjunction between them probably have a common explanation. The existing *Collomia mazama* in Oregon and its distant southern outpost, *C. rawsoniana*, probably represent the surviving remnants of a population system which was much more widespread in a former, moister age.

Collomia linearis is widespread in the forested region of western North America, ranging from the Pacific coast to the Great Lakes. Then after a gap of 1000 miles native colonies recur in New Brunswick and Quebec (Fig. 75c). Wherry (1936) remarks that in its eastern station the species is "in a region where many northwestern plants occur as relics of a former cross-continental distribution, disrupted by the last ice sheet."

Phlox austromontana occurs in the Great Basin and Colorado Plateau and again in the mountains of southern California. The disjunct area is separated from the main body of the species by the width of the Mojave Desert, a distance of 250 miles. The same disjunction separates some of the associated species of trees and shrubs. The narrowly endemic *Phlox dolichantha* of the San Bernardino Mountains in southern California is likewise separated from the related *P. stansburyi* of the Great Basin and Colorado Plateau by the gap of the Mojave Desert. It seems likely that the Phloxes in the southern California mountains are relics of a formerly more continuous distribution.

Gilia capitata tomentosa, a maritime entity, today has isolated inland populations on two mountains of central California which were surrounded by sea during the Pleistocene. Conversely, *G. capitata staminea*, an inland race of the San Joaquin Valley, occurs today on or near the California coast in several old tidal channels. Grant (1950) believes that these disjunct populations were stranded in their present positions by the withdrawal of the interior sea during the Pleistocene. *Gilia capitata capitata* occurs at present along the Pacific slope from British Columbia to San Francisco Bay. Several disjunct stations in eastern Oregon, eastern Washington and western Idaho may date back to Pleistocene times.

Gilia capillaris exists in a series of fairly closely spaced colonies through the Sierra Nevada and North Coast Range. This main area is separated from outlying areas by major and minor disjunctions, as shown in Figure 73. To the south isolated stations occur on Mt. Pinos and in the southern California mountains. To the north isolated populations occur in Chelan County, Washington. East of the Cascades a chain of disjunct stations stretches through the mountains of eastern Oregon to Idaho. Farther east in the Rocky Mountains several additional colonies are known, one from Clark County, Idaho, and another from Routt County, Colorado (Weber 1955). This latter disjunc-

Fig. 73. Distribution of Gilia capillaris in western North America.

tion represents a gap of 400 miles. The distribution pattern of *Gilia capillaris* appears to be a result of post-Pleistocene dissections of a previously more continuous range, although the information at hand is not sufficient to exclude alternative explanations.

Gilia laciniata, along with many other floristic elements, has a discontinuous distribution on the coastal plains and higher mountain zones of Peru, being absent from the very dry intermediate slopes of the Andes. Weberbauer (1939, 1945: 665) has suggested that during a pluvial stage in the Pleistocene these species had continuous areas, which have become broken up by climatic changes in later times.

Ipomopsis aggregata attenuata, a Rocky Mountain and Great Basin entity, occurs in one mountain pass in the Sierra Nevada under conditions which suggest that it is a Pleistocene relict there. *Ipomopsis congesta montana* occurs on isolated mountain peaks in the Sierra Nevada and Cascade Mountains. Some colonies are separated by distances of 75 miles. The broken area of this entity may also be of post-Pleistocene origin.

In the genus *Navarretia*, the area of *N. divaricata* in the Sierra Nevada and North Coast Range is separated by a prominent gap from an isolated colony on Mt. Pinos in south-central California. A similar disjunction of some 200 miles separates the northern branch of *Navarretia intertexta* (central California to British Columbia) from the southern branch in southern California. Whether the southern isolates are remnants of a Pleistocene distribution or products of long-range dispersal is not clear.

DISJUNCT AREAS IN NORTH AMERICA AND SOUTH AMERICA

A large contingent of species or species pairs belonging to many plant and animal groups have distribution areas in temperate North America and in temperate South America but are absent from the intervening tropical regions. This amphitropical distribution pattern, which has also been called bipolar, bifocal, antitropical or pan-temperate, has long been known to biogeographers and has been discussed by various authors[3].

It is well known, therefore, that South America harbors the close relatives of various North American species or vice versa. The interpretation of this distributional pattern, however, presents a problem. Two suggestions have been put forward to explain the presence of related populations in the widely distant areas. One possibility is that parallel forms have become differentiated

[3] See for example: Engler 1882 : 224; Gray and Hooker 1880; Bray 1898, 1900; Reiche 1907; Harshberger 1911; Pennell 1935: 590; Johnston 1940; DuRietz 1940; Penland 1941; Wulff 1943; Cain 1944, ch. 17; Campbell 1944; Weberbauer 1945; Axelrod 1950b; Stebbins 1950, ch. 14; Hubbs 1952; Good 1953.

independently on both sides of the equatorial region from a common ancestor in the tropics. The only other solution is trans-tropical migration.

The tropical forest is an age-old community and the tropical climate is an age-old environment. The northern and southern margins of the neotropical forest have shifted to lower latitudes during the ice ages and periods of aridity, but the tropical forest has always occupied the central area within the tropics. Temperate habitats in tropical latitudes are confined to the high mountains which, even in the present epoch of maximum uplift, are highly scattered and discontinuous. There has been, consequently, no extensive and continuous development of temperate habitats in the equatorial zone of the Western Hemisphere.

An ancestral population adapted to tropical conditions will be able to give rise to derivative populations on the northern and southern margins of the tropics by the known processes of evolutionary divergence. There is no particular difficulty in explaining the relationships between tropical and subtropical phylads now existing in northern and southern areas on the basis of a differentiation from a center of origin within the tropics.

The possibilities with regard to temperate phylads inhabiting disjunct areas on opposite sides of the tropics are more complex. The amphitropical distribution pattern of a temperate group may have arisen in three ways. The group may have originated from a temperate ancestor and in a temperate habitat in tropical latitudes. It may have diverged from a common tropical ancestor. Or the amphitropical group may have originated in the temperate zone of one hemisphere and migrated across the tropics to the opposite temperate region.

The first possibility is not likely. The evolutionary productions of the major temperate zones will on the average be able to hold their habitats against competitors originating in the minor temperate localities within the tropics, in the same way that continental faunas and floras usually prove superior in competition with insular biota. In a sample of temperate groups exhibiting an amphitropical distribution pattern only a small fraction of the cases, if any, would be expected to have a temperate center of origin within the tropics.

If temperate groups on opposite sides of the tropics have diverged independently from a tropical ancestor, their systematic relationships with one another will not be particularly close, since the evolutionary changes in passing from the ancestral form to either modern group will be considerable. The second possible explanation of an amphitropical distribution mentioned above will be invalid, therefore, if the disjunct temperate groups are more closely related to one another than they could possibly be to any common tropical ancestor. This is in fact the case for many amphitropical groups. The existence of close interrelationships between the northern and southern populations in a large sample of phylads favors, by the elimination of alternative explanations, the hypothesis of a temperate center of origin and subsequent trans-tropical migration, as applied to the majority of cases.

If we adopt the solution of trans-tropical migration we must next try to determine how the migration occurred. As noted above, the neotropical forest has continuously occupied a large central area in the tropical zone. This forest has formed a continuous barrier to the trans-tropical migration of temperate plants and animals. The crossing of this barrier may be accomplished in two ways: by a stepwise migration over trans-tropical land-bridges, or by rare feats of long-range dispersal.

A temperate species might spread by a series of steps from one temperate habitat to another within the tropical zone and thus in time reach the opposite hemisphere. It would find its stepping stones strung out in the mountain systems of Central America and equatorial South America. Prominent gaps exist in this land-bridge for a temperate species, however, particularly in the Isthmus of Panama, so that in order to cross the tropics some long jumps would be necessary.

The crossing of the tropics by a single hop would be what Simpson (1953) calls "sweepstakes dispersal". In this case the event has a very low probability of occurrence and may be long delayed. Given any probability of long-distance dispersal greater than zero, however, the migration to a new area across a formidable barrier may in time be achieved. Only one such event is required to explain the origin of a widely disjunct population.

We know that sweepstakes dispersal occurs. The colonization of the Hawaiian Islands and other oceanic islands has been by this method. For the Hawaiian Islands Simpson has calculated from data of Zimmerman that about five successful events of dispersal and establishment per 100,000 years would account for the angiospermous flora and insect fauna of these islands. A much lower rate of dispersal is indicated for the land birds and land snails of Hawaii, namely 0.3 and 0.5 colonizations per 100,000 years. No land mammal has succeeded in migrating to the Hawaiian Islands by natural means during five million years.

Whether a given species will be able to cross a given barrier to dispersal during a given period of time will depend on its inherent means of dispersal and on chance. The odds against sweepstakes dispersal of land mammals across large bodies of water are evidently very high. Water is not such a formidable barrier for flowering plants. Yet if the distances are great enough many plant species will be unable to cross them in the time available. An estimated 270 species of land plants migrated to Hawaii (Simpson 1953); many thousands of species existing together with the ancestors of these immigrants in the original floras did not reach Hawaii. The ability to cross water barriers is not unlimited in either land mammals or land plants.

Sweepstakes dispersal is partly a matter of chance. It is the chance of occurrence of a single event which has a very low probability of occurrence at all. As Axelrod (1952) has pointed out, this chance in the case of any plant group will be most likely to occur during geological periods when the climatic conditions are most favorable for that group. At such times the group will expand its area to the maximum extent. Disjunct populations in the direction of migration across the barrier may establish themselves as advance parties. The favorable habitats in the new area will be most numerous and widespread at this time. The number of seeds or vegetative propagules available for playing the game of sweepstakes dispersal will be greatest when conditions are at an optimum.

The two modes of trans-tropical migration, the dispersal in a series of steps over a land-bridge and the dispersal in one or a few long jumps, are accordingly not mutually exclusive. We have seen that some long jumps are necessary for any temperate group following along a trans-tropical land-bridge across Central America, because temperate habitats are not and have never been continuous on this land-bridge. On the other hand, the sweepstakes dispersal of temperate groups across the tropical zone is most likely to occur in periods of maximum contraction of the tropical forests, when the groups might have extended their areas into the tropics in various places.

The amphitropical species and species groups in the American hemisphere belong to a variety of communities. We find vicarious elements belonging to the cold temperate, alpine, cool temperate, steppe, mediterranean, desert and warm temperate regions of North America and South America. Many details are given by Harshberger (1911), Axelrod (1950b) and other authors cited in the footnote at the beginning of this section. I have listed below some representative amphitropical genera and species in each of the various regions. An up-to-date detailed analysis of the similarities and differences between the temperate and subtropical floras of North and South America, based on improved phytogeographic classifications and taking recent taxonomic findings into account, is badly needed. This would be a large but extraordinarily interesting study.

The regions with cold climates in North and South America have many floristic elements in common. A strong contingent of northern cold temperate genera occurs in the Andes (*Draba, Gentiana, Geum, Hieracium, Ranunculus, Ribes, Saxifraga, Spiraea, Vaccinium*)[4]. Disjunct distributions in the Rocky

---

[4] The family relationships of the genera mentioned in this and the next paragraphs are indicated here for the benefit of non-botanical readers.
Acacia (Leguminosae). *Acaena* (Rosaceae). *Agoseris* (Compositae). *Atriplex* (Chenopodiaceae). *Blennosperma* (Compositae). *Bromus* (Gramineae). *Caesalpinia* (Leguminosae). *Calandrinia* (Portulacaceae). *Calceolaria* (Scrophulariaceae). *Carex* (Cyperaceae). *Castilleja*

Mountains and Andes are found in several genera (*Castilleja, Pedicularis, Viola*) and in several species or superspecies (*Carex magellanica, Gentiana prostrata, Phleum alpinum, Primula farinosa, Saxifraga caespitosa*). The cool temperate rain forests of the Pacific Northwest and Magellanica have many floristic relationships as indicated by vicarious species pairs in *Carex, Epilobium, Festuca, Fragaria, Gayophytum, Libocedrus, Luzula, Osmorrhiza, Oxalis, Plantago, Poa,* and *Ranunculus.*

Some groups are common to the Great Basin and Patagonia (*Atriplex, Ephedra, Haplopappus, Lappula, Phacelia, Stipa*). The Great Plains and the Pampa are also vicarious, with equivalent members in *Bromus, Elymus,* and *Stipa.*

The warm temperate and subtropical floras of the two hemispheres likewise have many common components. In the mediterranean floras of cismontane California and central Chile paired species occur in such genera as *Agoseris, Blennosperma, Calandrinia, Chorizanthe, Clarkia, Madia, Mimulus, Phacelia,* and *Stachys*. There are identical or closely related species in the warm deserts of North America and South America. Paired species occur in *Acacia, Caesalpinia, Condalia, Ephedra,* and *Lycium,* while several characteristic species occur in both desert regions (*Larrea divaricata, Koeberlinia spinosa, Prosopis juliflora*). Still other groups are represented in both the Andes and the Mexican mountains (*Acaena, Calceolaria, Drimys, Fuchsia, Pernettya*).

Applying the rule that sweepstakes dispersal will occur most readily during the periods of maximum expansion of a population, it is evident that the components of the various communities which exhibit northern and southern disjunctions could not all have made the trans-tropical crossing in any single period. According to Axelrod (1950b, 1952) the best chances for the dispersal of subtropical woodland and forest elements occurred in the Miocene and early Pliocene; for steppe plants in the Pliocene; for boreal, alpine and cold-temperate species in the Pleistocene; and for warm desert species in the Pleistocene and Recent. No single hypothesis of dispersal will satisfactorily account for all the varied cases of amphitropical distribution.

Amphitropical distribution patterns are found in the Polemoniaceae. Their

(Scrophulariaceae). *Chorizanthe* (Polygonaceae). *Clarkia* (Onagraceae). *Condalia* (Rhamnaceae). *Draba* (Cruciferae). *Drimys* (Winteraceae).   *Elymus* (Gramineae). *Ephedra* (Ephedraceae). *Epilobium* (Onagraceae).   *Festuca* (Gramineae). *Fragaria* (Rosaceae). *Fuchsia* (Onagraceae). *Gayophytum* (Onagraceae). *Gentiana* (Gentianaceae). *Geum* (Rosaceae).   *Haplopappus* (Compositae). *Hieracium* (Compositae).   *Koeberlinia* (Capparidaceae). *Lappula* (Boraginaceae). *Larrea* (Zygophyllaceae). *Libocedrus* (Cupressaceae). *Luzula* (Juncaceae). *Lycium* (Solanaceae). *Madia* (Compositae). *Mimulus* (Scrophulariaceae).   *Osmorrhiza* (Umbelliferae). *Oxalis* (Oxalidaceae).   *Pedicularis* (Scrophulariaceae). *Pernettya* (Ericaceae). *Phacelia* (Hydrophyllaceae). *Phleum* (Gramineae). *Plantago* (Plantaginaceae). *Poa* (Gramineae). *Primula* (Primulaceae). *Prosopis* (Leguminosae).   *Ranunculus* (Ranunculaceae). *Ribes* (Saxifragaceae). *Saxifraga* (Saxifragaceae). *Spiraea* (Rosaceae). *Stachys* (Labiatae). *Stipa* (Gramineae).   *Vaccinium* (Eriaceae). *Viola* (Violaceae).

occurrence in this family is a part of a wider problem of floristic and faunistic evolution. With the foregoing discussion of general principles as a background, it may be worthwhile now to see what light the evidence from the Polemoniaceae can shed on this fascinating biogeographical problem.

The polemoniaceous flora of South America is a peculiar mixture of primitive and advanced forms. In the tropical regions we have representatives of *Cobaea, Cantua* and *Loeselia*, all of which possess relatively primitive features. In the temperate parts of South America are found *Cantua* and *Huthia*, among the primitive forms, as well as a variety of herbaceous types with advanced characters.

The Cobaeas of South America have tropical American relationships. *Loeselia glandulosa* in South America has a Central American or Mexican center of distribution. The Cantua tribe, as we have already pointed out elsewhere, has probably diverged from a tropical American ancestor. It occupies much the same position in relation to some primitive tropical ancestor as does the Bonplandia tribe on the northern edge of the tropics.

With regard to the tropical Polemoniaceae, therefore, the hypothesis of a parallel evolution of different northern and southern populations from a common tropical ancestor provides the simplest and most satisfactory explanation of the facts. In *Loeselia glandulosa*, which as a species spans the tropics, the parallel differentiation has not gone beyond the level of races (Fig. 74a). The divergence between the northern and southern branches of *Cobaea* is on the species level (Fig. 74b). The divergence separating the Mexican *Bonplandia* from the Andean *Cantua* and *Huthia* has reached the generic or tribal level (Fig. 74c). But in every case this radiation into parallel northern and southern stocks can have taken place from a common point of departure within the tropics.

Let us next inquire whether this same simple explanation will suffice in the case of the herbaceous Polemoniaceae of temperate South America. The twelve species of austral herbaceous Polemoniaceae which can be recognized taxonomically at the present time are listed in Table 13.

These twelve species are in all cases related to particular North American species, although a great deal remains to be learned about the details of the relationships. With few exceptions the South American species are not as closely related to one another as they are to different North American species. These facts rule out the possibility of an independent differentiation of the northern and southern forms from an ancestral tropical plexus and point instead to a temperate origin and trans-tropical migration.

Certain additional facts indicate that this temperate center of origin lay in North America and not in South America. Table 13 shows that the twelve austral species are distributed among nine separate phylads. All of these phylads occur in North America; most of them are richly differentiated in North America and only represented by one or two species in the southern hemisphere; while no amphitropical phylad is better developed in South America than in North America. Furthermore, the South American species represent in most cases derived forms within their respective phylads. Only

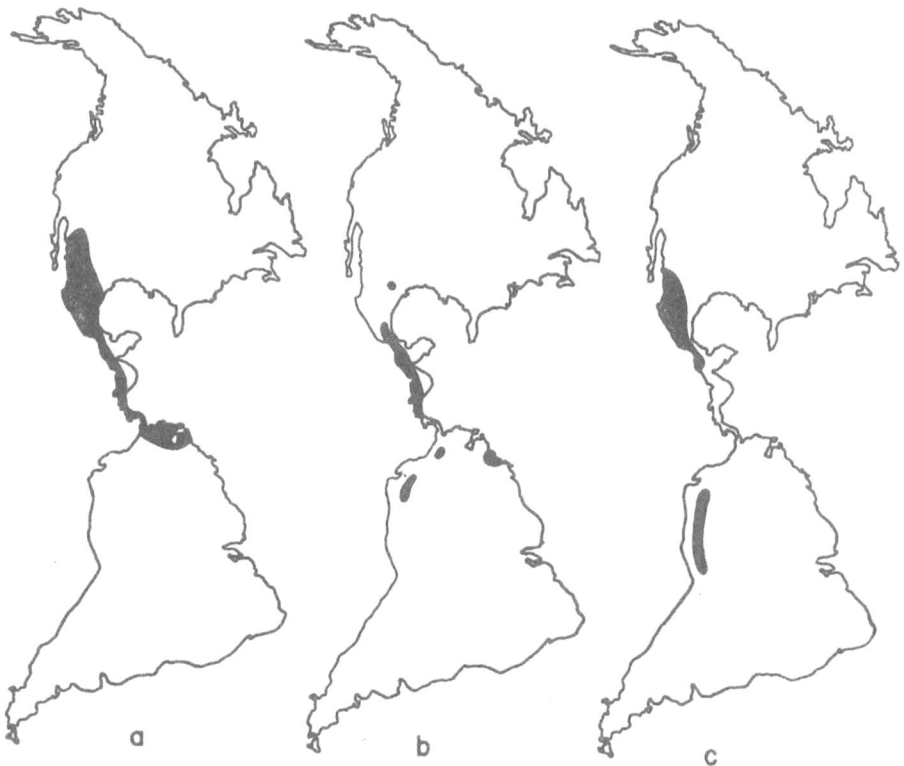

Fig. 74. Trans-tropical distribution patterns in the Polemoniaceae.
a) Loeselia glandulosa. *b*) Cobaea. *c*) Bonplandia (north) and Cantueae (south).

one is at all primitive for its phylad; the other eleven are among the most highly reduced forms of their genus or section. The exceptional twelfth species, *Gilia foetida*, is a cross-pollinating perennial herb, whereas all the other species are annuals, often quite reduced in size and general morphology, with a self-pollinating breeding system. These facts point to the conclusion that the herbaceous South American members of the Phlox family have North American floristic affinities and have been derived from northern centers of origin.

The distribution areas of the amphitropical species and species pairs, so far as they are known at present, are shown in Figures 75, 76 and 77. It was necessary to draw many small areas larger than they actually are in order to make them stand out on the large-scale maps.

The vicarious races and species frequently occupy generally equivalent habitats in their northern and southern areas. Thus the amphitropical representatives of *Collomia* sect. *Collomia, Microsteris gracilis,* and the *Gilia*

Table 13. Systematic Relationships of the South American Herbaceous Polemoniaceae

Genus and Section	South American Representatives	Closely related North American Species	No. of North American Species in the same Group
Polemonium sect. Polemoniastrum	P. micranthum	P. micranthum	1
Collomia sect. Collomia	C. biflora C. cavanillesii	C. grandiflora C. linearis	6
Microsteris	M. gracilis	M. gracilis	1
Gilia sect. Giliastrum	G. foetida G. glutinosa	G. rigidula G. palmeri	5
Gilia sect. Gilia	G. laciniata G. valdiviensis	G. nevinii G. millefoliata	7
Gilia sect. Arachnion	G. crassifolia	G. sinuata	17
Ipomopsis sect. Microgilia	I. gossypifera	I. pumila	10
Navarretia sect. Navarretia	N. involucrata	N. minima	7
Linanthus sect. Dactylophyllum	L. pusillus	L. pygmaeus	11

millefoliata-valdiviensis group occur in both the Pacific Northwest and Magellanica; *Microsteris gracilis* occurs in both the Rocky Mountains and Andes; the Great Basin and Patagonia are both inhabited by *Polemonium micranthum, Microsteris gracilis, Gilia sinuata-crassifolia,* and *Navarretia minima-involucrata.* In the respective mediterranean communities of the two

Fig. 75. Amphitropical distribution patterns in the Polemonium Tribe.
a) Polemonium micranthum. *b)* Collomia grandiflora (north) and C. biflora (south). *c.* Collomia linearis (north) and C. cavanillesii (south). *d)* Microsteris gracilis. (Recent range extensions of Collomia linearis in the eastern United States and of this species and Microsteris gracilis into the Yukon resulting from human influences are not shown.)

Fig. 76. Amphitropical distribution patterns in the Gilia Tribe.
The first-listed species is northern, the second-listed southern.
a) Gilia rigidula-foetida. *b)* Gilia palmeri-glutinosa. *c)* Gilia millefoliata (in north); G. valdiviensis and G. laciniata erecta (in south). *d)* Gilia nevinii-laciniata.

Fig. 77. Amphitropical distribution patterns in the Gilia Tribe (cont.)
The first-listed species is northern, the second-listed southern.
a) Gilia sinuata-crassifolia. *b)* Ipomopsis pumila-gossypifera. *c)* Navarretia minima-involucrata. *d)* Linanthus pygmaeus-pusillus.

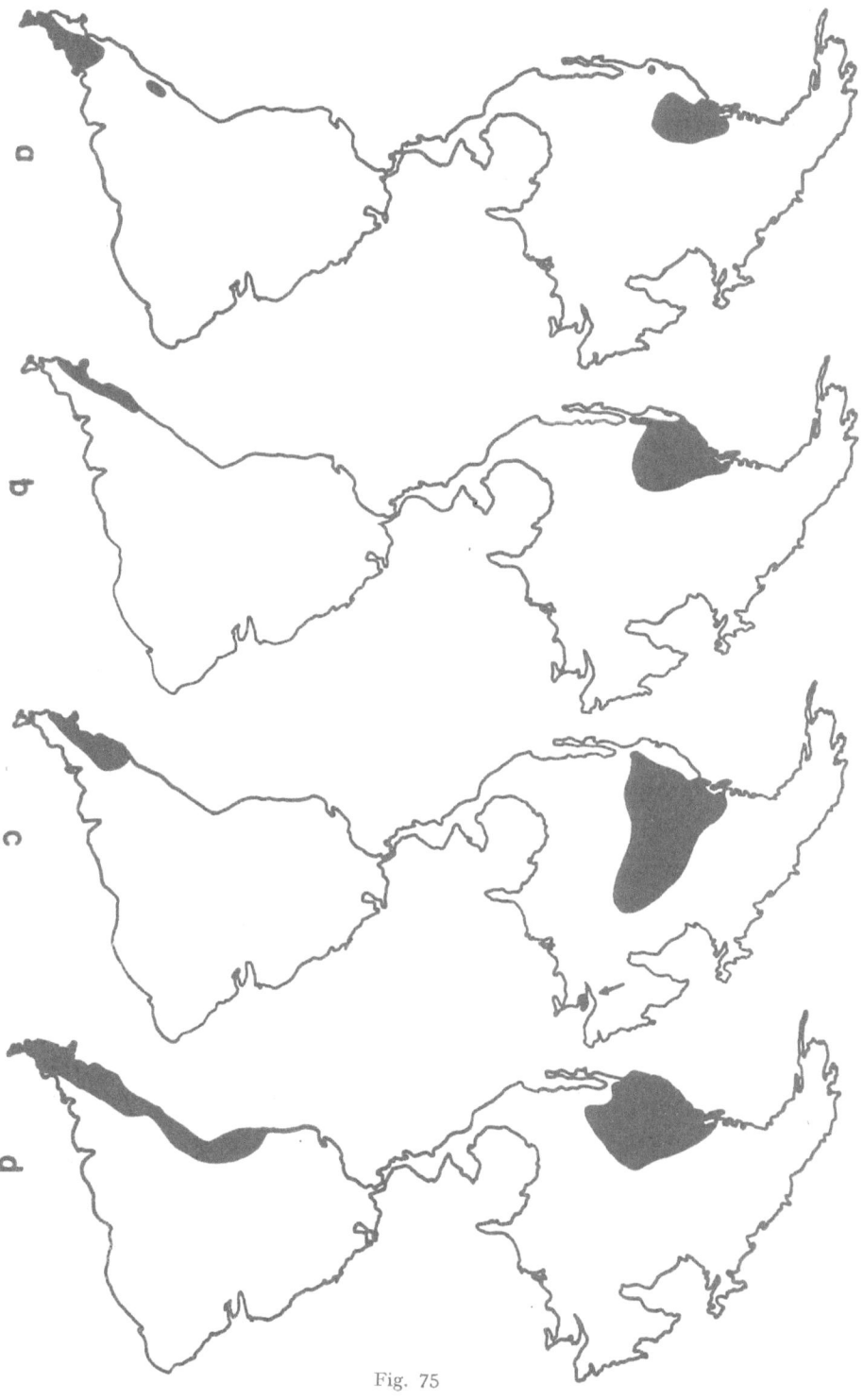

Fig. 75

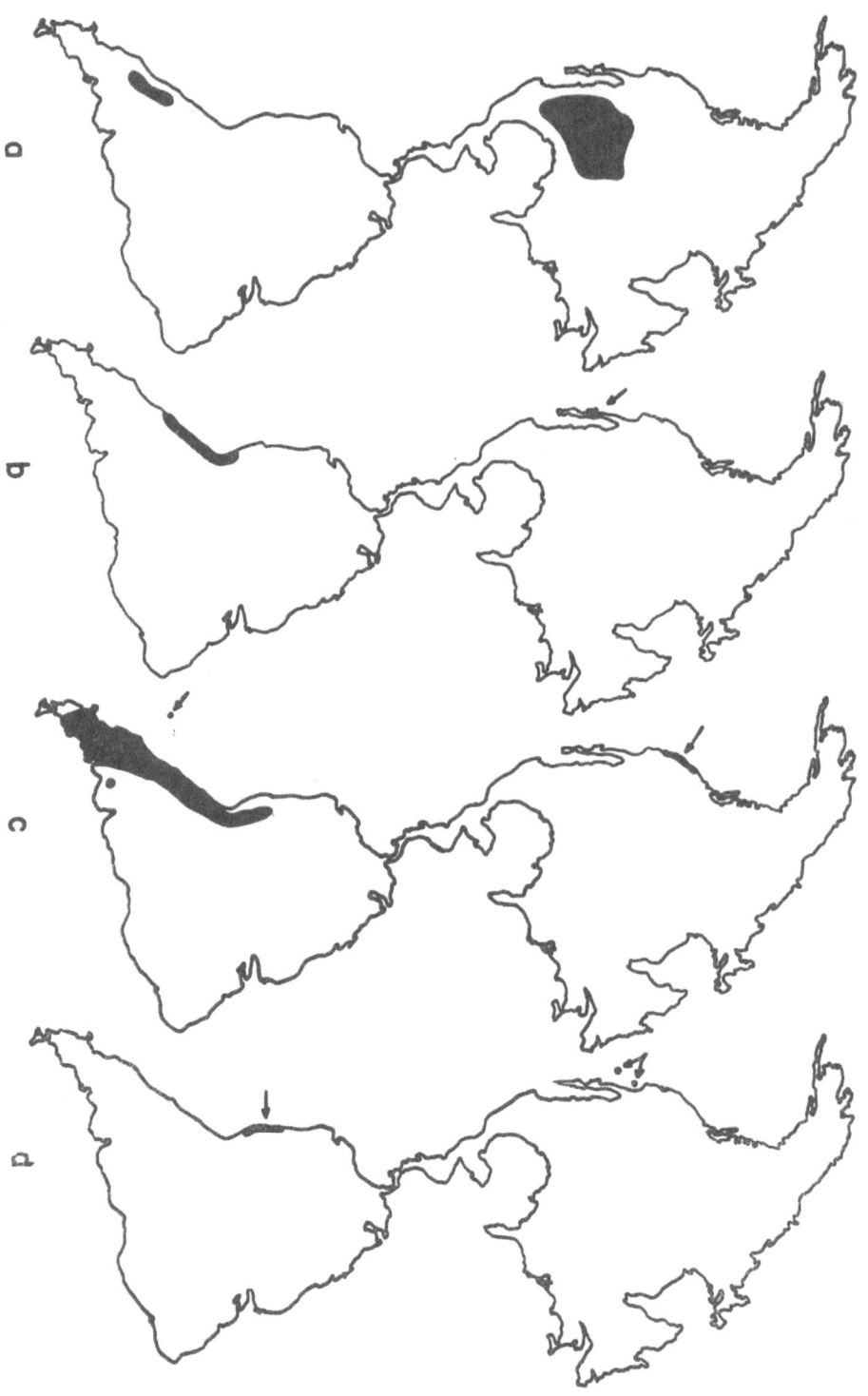

Fig. 76

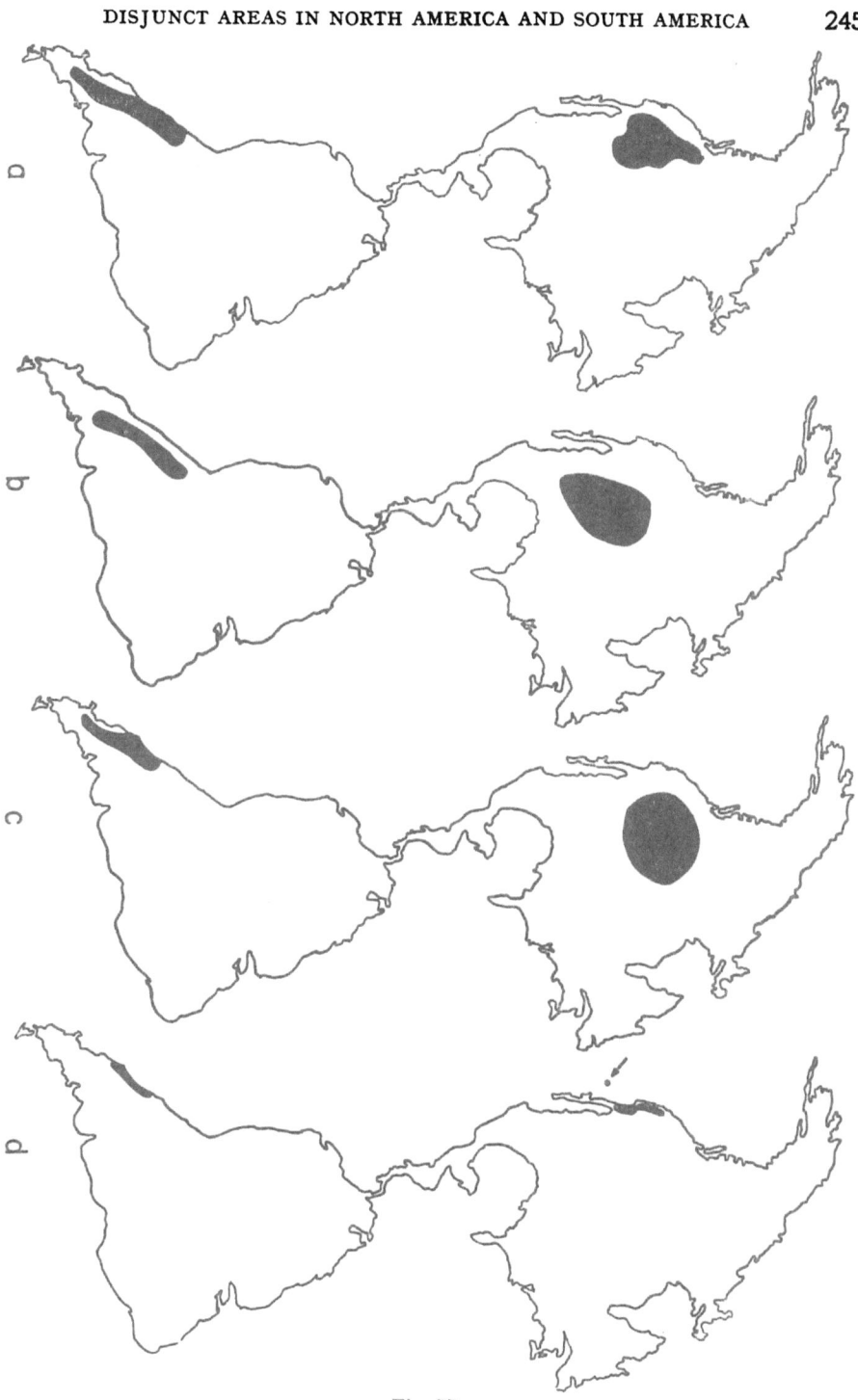

Fig. 77

continents we find *Microsteris gracilis, Gilia millefoliata-valdiviensis,* and *Linanthus pygmaeus-pusillus.* The coastal deserts are inhabited by the paired species *Gilia palmeri-glutinosa* and *Gilia nevinii-laciniata;* and the interior deserts by *Gilia rigidula-foetida, Ipomopsis pumila-gossypifera,* and *Navarretia minima-involucrata.*

The northern and southern areas are seen in Figures 75, 76 and 77 to be separated by the width of the tropics, a disjunction of thousands of miles. The range of *Polemonium micranthum,* for example, is disrupted by a gap 5500 miles wide (Fig. 75a). All the gaps shown in Figures 75 to 77 are between 4000 and 5500 miles across.

The ability of the Polemoniaceae to cross minor dispersal barriers is attested to by the occurrence of representatives of the family on many islands separated by scores or hundreds of miles from the nearest land. Thus *Polemonium boreale* occurs in Greenland, Spitsbergen and Novaya Zemlya; *Polemonium caeruleum* in England; *Gilia valdiviensis* on Masafuera in the Juan Fernandez Islands; *Gilia nevinii* and *Linanthus pygmaeus* on Guadalupe Island off Baja California; and various species on the coastal California islands.

On the other hand, the Polemoniaceae are absent from the native floras of Africa and Australia, though *Phlox drummondii* has spread on a limited scale in South Africa and *Navarretia squarrosa* on a wider scale in Australia since their introduction into these continents by man. These or other species of Polemoniaceae might have been able to colonize Africa and Australia in prehistoric times except for the dispersal barrier of approximately 8500 miles. Many of the Polemoniaceae, therefore, can jump across gaps scores or hundreds of miles wide, a few have been able to migrate several thousand miles across the tropics in one or several successive hops, but none has the proven ability of crossing distances greater than about 5500 miles.

By what means of dispersal have the few occasional long hops, measured in many hundreds or a few thousands of miles, taken place? It is a significant fact that every one of the twelve herbaceous species in South America has mucilaginous seeds, even though this represents in some cases an exceptional condition for the genus or section. Such seeds when damp develop mucilaginous spiracles and/or slime cells on the outer coat which cause them to adhere to even the slickest surfaces. Among the objects to which the seeds might occasionally become attached would be the feet or feathers of birds migrating between North and South America.

The disjunct distribution areas shown in Figures 75 to 77 correspond to known migration routes of birds. Many shorebirds such as the Wilson Phalarope, Spotted Sandpiper, Killdeer and Snowy Plover pass regularly from western North America to coastal Peru, Chile and Patagonia (Cooke 1910). These birds inhabit the open terrestrial plant formations where many of the

Polemoniaceae are found. It is difficult to imagine a method of long-range dispersal better fitted to explain the disjunct amphitropical areas of the Polemoniaceae than transport by migratory birds in general and by shore-birds in particular.

For a number of years I thought this idea was original. In searching the literature, however, it turned out that Bray in a rarely cited paper had expressed it over a half century earlier (Bray 1898). Referring to the amphi-tropical Polemoniaceae, he writes: "The method of distribution is of interest here. Commonly in the Polemoniaceae the seed is furnished with a layer of cells whose walls become mucilaginous by contact with water, expelling forcibly the spirally thickened hair-like processes which cause the seed to adhere firmly to moistened objects (e.g., feet of birds) and thus secure trans-portation. So far as I was able to examine, all the South American species of [herbaceous] Polemoniaceae possess these mucilaginous seeds." Bray points out that mucilaginous seeds are found in some other amphitropical groups such as *Fagonia* and *Peganum* of the Zygophyllaceae.

A second important characteristic of the herbaceous Polemoniaceae in South America which is apparently related to their migratory potential is the fact that all but one of the species are autogamous. The significance of the strong preponderance of self-pollinators in South America may be that chance introductions of these into a new area are more likely to set seeds and reproduce there than are outcrossing types. The cross-pollinating species adapted to a certain set of normal pollinators might be unable to survive beyond the first generation in a strange land where the proper pollinating agents are absent. Furthermore, the introduction of a single seed would suffice to found a new population in a plant with an autogamous system of reproduction, whereas for obligatorily outcrossing plants the foundation of a new population would depend on the simultaneous or nearly simultaneous introduction of at least two seeds. In a plant group playing for sweepstakes dispersal the type of breeding system will often prove to be a critical factor. Baker (1953, 1955) has arrived at the same conclusions on the basis of his studies of the Plumbaginaceae and other families.

It should be noted, finally, that the North American relatives of the South American species are in a majority of the cases wide ranging. The presence of large numbers of individuals scattered over large territories would increase the probabilities of dispersal. The few North American representatives of amphi-tropical groups that have narrow distribution areas are all maritime. This is the case in *Gilia palmeri*, *G. millefoliata*, *G. nevinii*, and *Linanthus pygmaeus*. Their seeds might be picked up and carried to South America by oceanic birds.

The question whether the migration to South America occurred in a series of stages or in a single long jump cannot be answered with the evidence available. If way-stations existed for some of the species in the high mountains of Mexico, Guatemala, Colombia, and Ecuador we might be justified in assuming a stepwise migration. The absence of such way-stations through the tropics at the present time, however, does not entitle us to reject the possibility of a migration by leaps and bounds. Negative evidence is a dangerous grounds for conjecture in phytogeography. We can only note that a dispersal by stages is not the only, or even perhaps the most likely, possibility. If the ancestors of the South American Gilieae and Polemonieae arrived in their new homeland on the feet or feathers of migratory birds they could easily have crossed the tropical zone in a single hop.

We may assume, following Axelrod, that the chances of dispersal from North America to South America were greatest during periods when the climatic conditions were most favorable for each immigrating species. Judging by the probable age of the communities with which the herbaceous Polemoniaceae of South America are now associated, the immigration probably took place in relatively recent times. A few species, as for example *Gilia foetida*, might have arrived in South America as early as the Pliocene. For the majority of the amphitropical Polemoniaceae, however, and perhaps for all of them, the conditions for dispersal and establishment would have been optimal at various times during the Pleistocene and Recent periods.

CHAPTER IX

GENERAL CONCLUSIONS

The present work strives to approach a conception of systematic botany which was well expressed by Danser. "Many false conceptions have arisen in systematics because living beings were treated as objects which the scientist proposed to classify. Once for all attention must be drawn to the fact that the systematist never classifies objects but life-cycles ... The objects which we use as material [in museums] ... are nothing but an often inevitable substitute for the living beings represented by them. The missing links of the life-cycle in these objects must be completed as far as possible by the know-ledge of the systematist if his systematics is to preserve its scientific character. Even the structure of the full-grown stage, on which, in many groups, systematics is almost exclusively based, must, however great at times its importance, be only looked upon as the representation of the life-cycle.

"This conception of systematics makes it necessary that various matters which are usually considered to belong to other sciences should not be neglected in systematics as soon as they increase our knowledge of the life-cycle of living beings and are useful for the science of classification, such as certain parts of ecology and physiology ..., of genetics ..., and of chemistry ...

"In this connection attention should be drawn to the widely-spread yet erroneous idea that a systematist who, in order to solve a systematic problem, engages in a genetic, cytological, physiological or chemical research, or who even makes use of genetic, cytological, physiological or chemical methods in his research, would exceed the limits of systematics ..." (Danser 1950: 118–119).

A generalized classification of the Phlox family is proposed in this work. An outline of the system is given below together with a citation of a few well known species for each genus and section.

A. Tribe Cobaeeae.

    I. Cobaea.

1. Sect. Cobaea. (*C. scandens.*)
2. Sect Aschersoniophila. (*C. aschersoniana.*)
3. Sect. Rosenbergia. (*C. hookeriana, C. penduliflora.*)

B. Tribe Cantueae.
    II. Cantua. (*C. buxifolia, C. pyrifolia.*)
    III. Huthia. (*H. coerulea.*)

C. Tribe Bonplandieae.
    IV. Bonplandia. (*B. geminiflora.*)
    V. Loeselia.
        1. Sect. Loeselia. (*L. glandulosa, L. mexicana.*)
        2. Sect. Glumiselia. (*L. grandiflora.*)

D. Tribe Polemonieae.
    VI. Polemonium.
        1. Sect. Polemonium. (*P. boreale, P. caeruleum, P. pauciflorum.*)
        2. Sect. Melliosma. (*P. confertum, P. viscosum.*)
        3. Sect. Polemoniastrum. (*P. micranthum.*)

    VII. Allophyllum. (*A. gilioides, A. glutinosum.*)

    VIII. Collomia.
        1. Sect. Collomiastrum. (*C. debilis, C. larseni.*)
        2. Sect. Courtoisia. (*C. heterophylla.*)
        3. Sect. Collomia. (*C. biflora, C. grandiflora, C. linearis.*)

    IX. Gymnosteris. (*G. nudicaulis.*)

    X. Phlox.
        1. Sect. Phlox. (*P. adsurgens, P. carolina, P. glaberrima, P. longifolia, P. ovata, P. paniculata, P. stansburyi, P. subulata.*)
        2. Sect. Divaricatae. (*P. amoena, P. divaricata, P. drummondii, P. nivalis.*)
        3. Sect. Occidentales. (*P. caespitosa, P. diffusa, P. douglasii, P. hoodii, P. sibirica.*)
    XI. Microsteris. (*M. gracilis.*)

E. Tribe Gilieae.
    XII. Gilia.
        1. Sect. Giliastrum. (*G. filiformis, G. foetida, G. incisa, G.rigidula.*)
        2. Sect Giliandra. (*G. leptomeria, G. pinnatifida, G. subnuda.*)

3. Sect. Gilia. (*G. achilleafolia, G. capitata, G. laciniata, G. millefoliata, G. tricolor.*)
4. Sect. Arachnion. (*G. cana, G. crassifolia, G. inconspicua, G. latiflora, G. sinuata, G. tenuiflora.*)
5. Sect. Saltugilia. (*G. capillaris, G. leptalea, G. splendens, G. G. stellata.*)

XIII. Ipomopsis.
    1. Sect. Phloganthea. (*I. gloriosa, I. multiflora.*)
    2. Sect. Ipomopsis. (*I. aggregata, I. longiflora, I. rubra.*)
    3. Sect. Microgilia. (*I. congesta, I. pumila, I. spicata.*)

XIV. Eriastrum. (*E. densifolium, E. diffusum, E. virgatum.*)

XV. Langloisia. (*L. matthewsii, L. punctata.*)

XVI. Navarretia.
    1. Sect. Aegochloa. (*N. atractyloides, N. squarrosa.*)
    2. Sect. Masonia. (*N. breweri, N. divaricata.*)
    3. Sect. Mitracarpium. (*N. intertexta, N. tagetina.*)
    4. Sect. Navarretia. (*N. involucrata, N. minima.*)

XVII. Leptodactylon. (*L. californicum, L. pungens.*)

XVIII. Linanthus.
    1. Sect. Siphonella. (*L. nuttallii.*)
    2. Sect. Pacificus. (*L. grandiflorus.*)
    3. Sect. Leptosiphon. (*L. androsaceus, L. bicolor, L. montanus.*)
    4. Sect. Dactylophyllum. (*L. aureus, L. harknessii, L. pusillus.*)
    5. Sect. Linanthus. (*L. bigelovii, L. dichotomus.*)
    6. Sect. Dianthoides. (*L. dianthiflorus, L. parryae.*)

The character combinations which support the above system of classification are summarized in Chapter 4, and the taxonomic interpretation of these characters is discussed in Chapter 3. Identification of a specimen as to tribe, genus and section will be facilitated by the use of the tables of characters in Chapter 4.

The nomenclature of the family is reviewed in Chapter 5. Complete lists of generic, infra-generic and specific names hitherto proposed in the Polemoniaceae are given at the end of Chapter 5. The sections *Glumiselia* of *Loeselia* and *Masonia* of *Navarretia* are new; five additional sectional names in *Linanthus* and the section *Aegochloa* of *Navarretia* represent new combinations; and two specific names in *Gilia* and *Linanthus* are published as new combinations.

The distribution of the 316 species among the 18 genera is not random. There are two large genera with 56 to 61 species (*Gilia* and *Phlox* respectively), seven medium-sized genera with 14 to 37 species, five small genera with 5 to 9 species, and four very small genera with one or two species each. This pattern, following a suggestion of Stebbins (1951), may mean that a series of fortunate circumstances leading to an outstanding evolutionary success occurred only rarely in the history of the family. The rare combination of favorable conditions was achieved in the two large genera, *Phlox* and *Gilia*, which incidentally are segregated into different centers of abundance. "Some of these conditions would be fulfilled in a large number of instances, so that more genera of moderate size could come into being. And the number of combination of characters sufficiently adaptive to permit their bearers to survive in a single ecological niche, or for a short period of geological time, is relatively large, so that a large number of small genera would be expected in any family" (Stebbins 1951).

Fig. 78. Evolutionary trends in several morphological characters in the Phlox family. Living representatives of the different stages shown, not forming a phylogenetic series, are as follows.

CHROMOSOMES. – *minute*: tropical genera. *large*: Polemonium, Ipomopsis. *medium*: Collomia, Gilia. *small*: Linanthus. *median centromeres*: Polemonium, Gilia. *subterminal centromeres*: Phlox, Linanthus. *9 pairs*: Loeselia, Polemonium, Gilia. *8 pairs*: Collomia, Gilia sect. Giliandra. *7 pairs*: Phlox, Ipomopsis. *6 pairs*: Gymnosteris.

VEGETATIVE BODY. – *embryo colorless*: tropical genera. *embryo green*: most temperate genera. *tree*: Cantua. *shrub*: Cantua. *subshrub*: Leptodactylon. *perennial herb*: Polemonium. *biennial*: Ipomopsis sect. Ipomopsis. *annual*: Gilia. *vine*: Cobaea. *foliage all leafy*: tropical genera. *foliage leafy and bracteate*: Gilia. *foliage bracteate*: Gymnosteris. *leaves alternate*: most genera. *leaves alternate and opposite*: Microsteris. *leaves opposite*: Phlox, Linanthus.

LEAF. – *a–d*) Polemonium tribe. *a'-d'*) Cobaea and Cantua tribes. *a"–c"*) Gilia tribe. a) Polemonium foliosissimum. b) Collomia rawsoniana. c) Phlox glaberrima. d) Phlox covillei. a') Cobaea scandens. b') Huthia coerulea. c') Cantua buxifolia. d') Cantua pyrifolia. a") Ipomopsis aggregata aggregata. b") Ipomopsis tenuifolia. c") Linanthus grandiflorus. (all × 1/2.)

CALYX. – a) Cobaea scandens. b) Polemonium pectinatum. c) Collomia grandiflora. d) Phlox stolonifera. e) Eriastrum pluriflorum.

COROLLA. – *a–e*) Polemonium tribe. *a'–e'*) Gilia tribe. a) Polemonium carneum. b) Polemonium viscosum. c) Polemonium pauciflorum. d) Phlox dolichantha. e) Microsteris gracilis. a') Gilia rigidula. b') Gilia capitata staminea. c') Langloisia matthewsii. d') Linanthus bicolor. e') Linanthus harknessii. (all life-size.)

STAMENS. – a) Cobaea scandens. b) Polemonium caeruleum occidentale. c) Collomia larsenii. d) Phlox stolonifera. e) Collomia mazama. f) Navarretia pleiantha. (semi-diagrammatic, not to scale.)

CAPSULE. – a) Cobaea scandens. b) Cantua pyrifolia. c) Loeselia mexicana. d) Gilia capitata capitata. e) Navarretia intertexta. (a × 1-1/4, b–e × 2-1/2; number of locules and mode of dehiscence shown in accompanying diagrams.)

CHROMOSOMES

Size	minute large → medium → small
Position of Centromere	mostly median → mostly subterminal
Number of Pairs	9 → 8 → 7 → 6

VEGETATIVE BODY

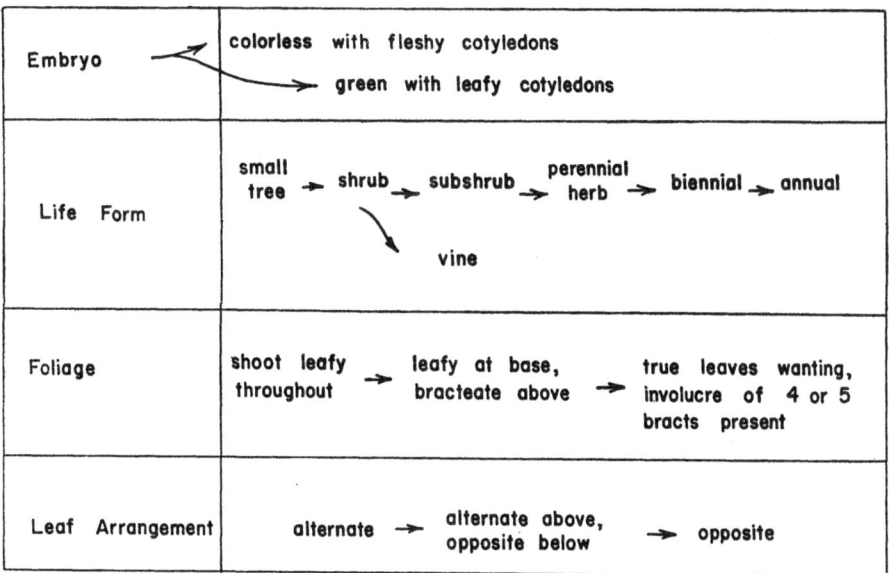

Embryo	colorless with fleshy cotyledons green with leafy cotyledons
Life Form	small tree → shrub → subshrub → perennial herb → biennial → annual vine
Foliage	shoot leafy throughout → leafy at base, bracteate above → true leaves wanting, involucre of 4 or 5 bracts present
Leaf Arrangement	alternate → alternate above, opposite below → opposite

Fig. 78. Evolutionary trends in several morphological characters. For identification of the stages see the facing page.

LEAF   (all X 1/2)

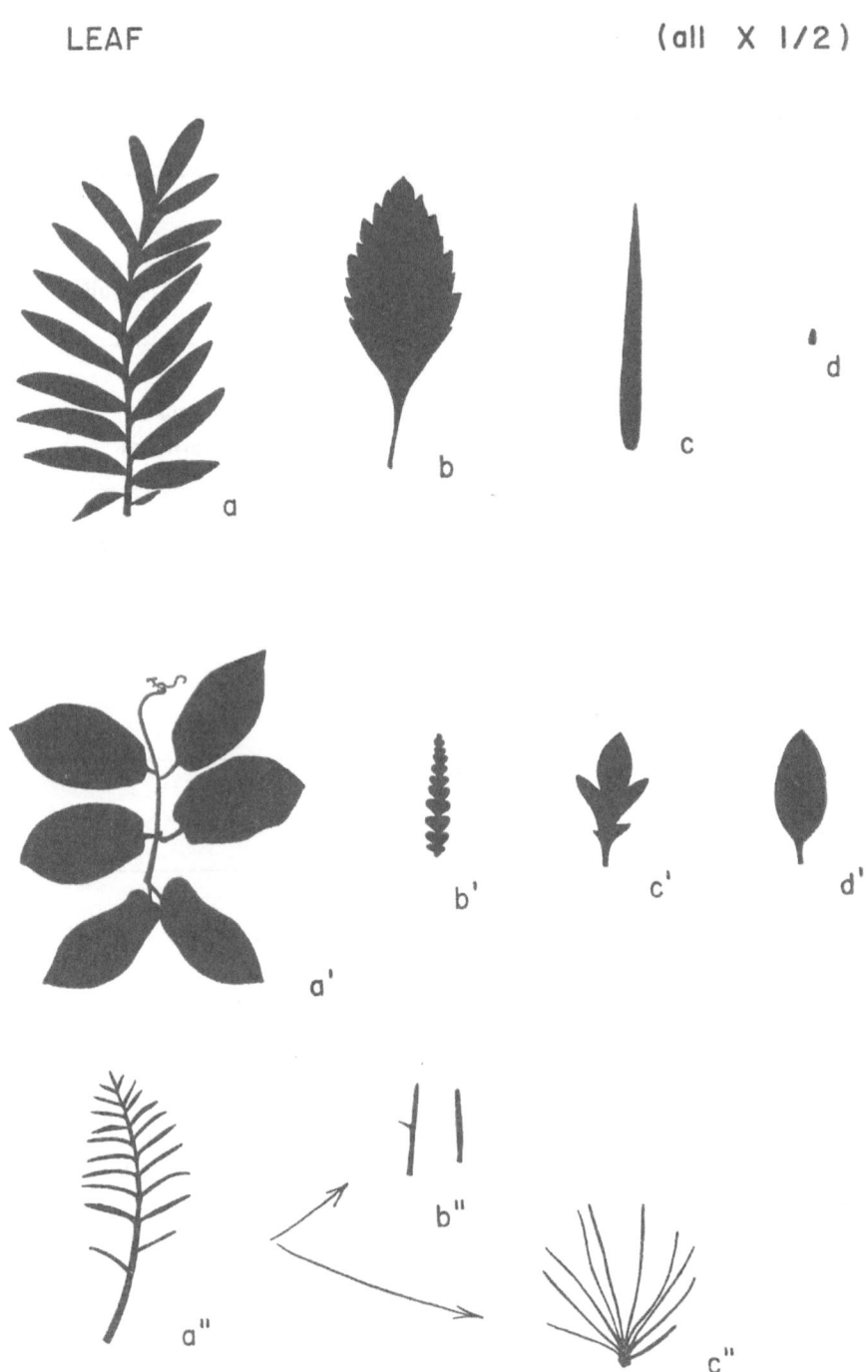

Fig. 78. continued.

CALYX $\qquad$ (a X I, b-e X 2)

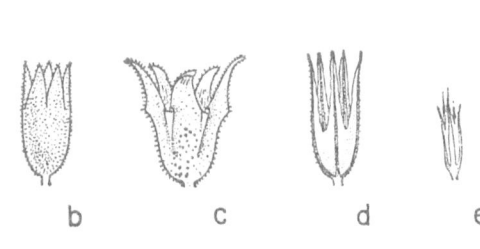

a $\qquad$ b $\qquad$ c $\qquad$ d $\qquad$ e

COROLLA $\qquad$ (all X I)

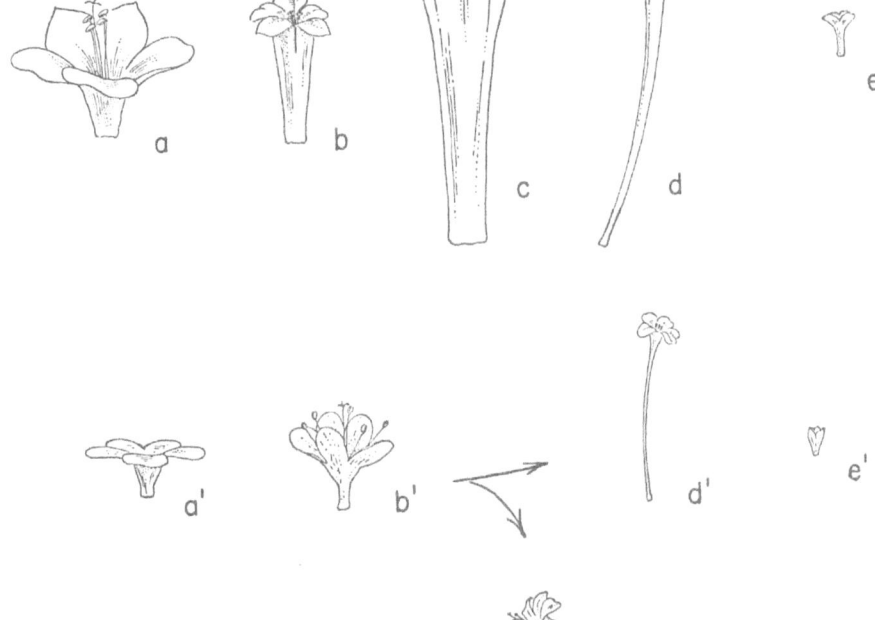

Fig. 78. continued.

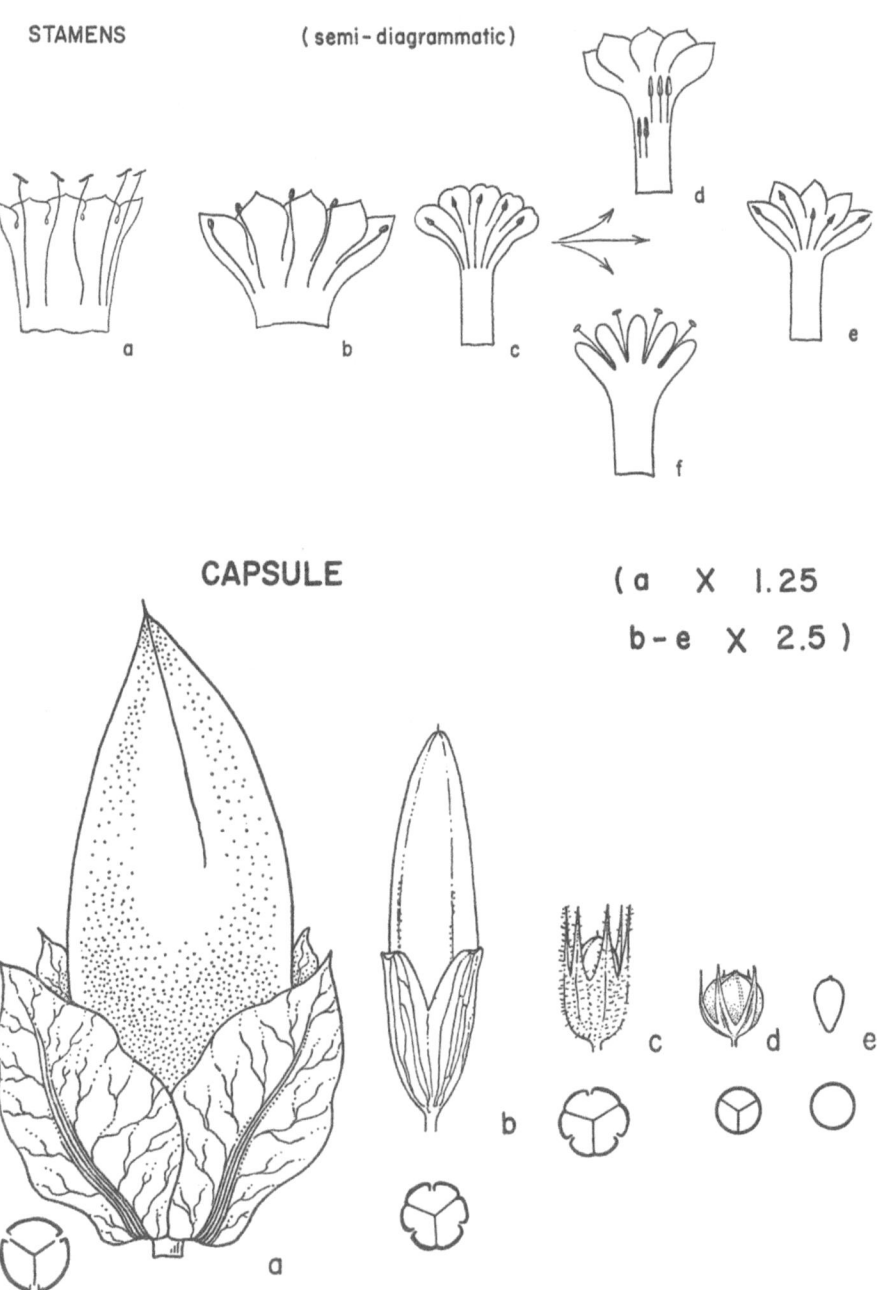

STAMENS   (semi - diagrammatic)

CAPSULE   (a   X   1.25

b - e   X   2.5)

Fig. 78. continued.

A complete summary of chromosome numbers in the Polemoniaceae is given in Chapter 6. This summary lists the numbers for 132 species, of which 50 are newly reported and 82 are brought together from the literature. Many of the latter, being recorded in specialized taxonomic papers, have been overlooked in recent general compendia of chromosome numbers.

Twenty-three percent of the species counted are polyploid. These polyploids are irregularly distributed throughout the various genera. Polyploidy is most common in *Cantua* and *Cobaea*, fairly frequent in *Gilia* and *Collomia*, uncommon in *Phlox*, *Polemonium* and *Ipomopsis*, and either absent or undiscovered in the remaining genera. Most of the polyploids are tetraploids; hexaploidy is found in *Cantua*, *Cobaea* and *Gilia*; and octoploidy also oocurs in *Gilia*.

The basic number for the family as a whole is X = 9. A descending aneuploid series from X = 9 to X = 8, 7 and in one instance 6 has occurred independently in the two temperate tribes, Polemonieae and Gilieae. For reasons which are not clear the number N = 8 is much less frequent than the neighboring numbers N = 9 and 7.

Many morphological characters exhibit a wide range of variation within the family from a condition approximating that found in primitive Dicotyledons to one or more derived conditions. The evolutionary trends for several of these characters are summarized graphically in Figure 78.

No single living group possesses a monopoly of primitive features or of advanced features. The different morphological characters have evidently evolved at different rates within any given phyletic line. The character phylogenies shown in Figure 78 are to a large extent independent of the phylogenetic relationships between actual groups. Nevertheless, some phylads possess a relatively large number of characteristics like those believed to be present in the ancestral Polemoniaceae, while other phylads have retained very few ancestral characters. This means that some existing phylads have diverged with respect to most of their characters very little, and others a long way, from the ancestral condition of the family. The former may be regarded as relatively primitive groups, the latter as advanced groups (cf. Sporne 1948).

The phylogenetic development of the Phlox family cannot be represented diagrammatically by a branching tree, since this development has involved the anastomosis as well as the divergence of phyletic lines (cf. Fig. 66). The reticulate pattern of evolution seems to be a very common one in higher plants. The general degree of advancement of the different living phylads and their affinities to one another can, however, be represented satisfactorily against a target background (Sporne 1956). Figure 79 is an effort to portray

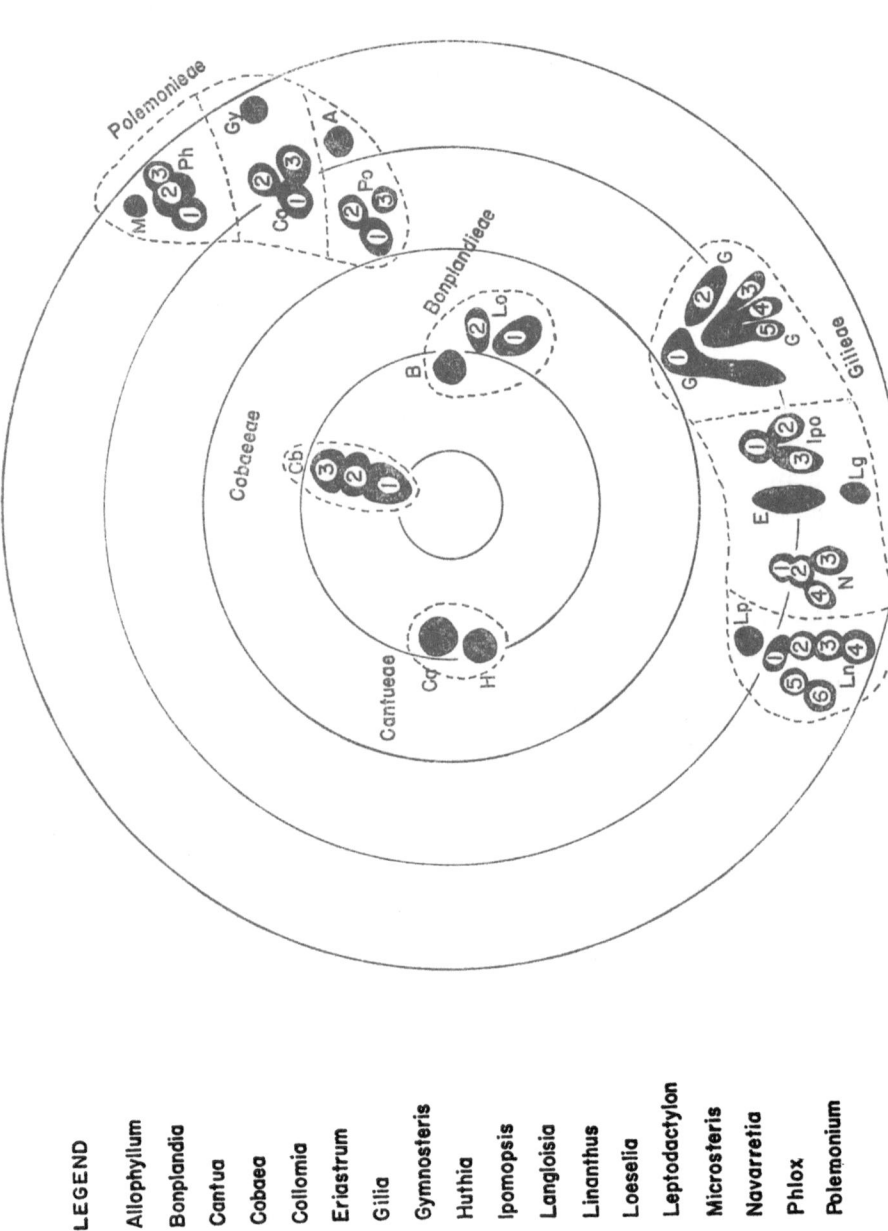

Fig. 79. Phylogenetic relationships in the Phlox family.

Relative advancement is represented by position on the radius of the target, and affinities by proximity of the groups to one another. Tribes and subtribal groupings are outlined by dashed lines; genera are shown in solid black. Within the genera the code numbers for sections are the asasme those used in the text.

LEGEND

A Allophyllum

B Bonplandia

Ca Cantua

Cb Cobaea

Co Collomia

E Eriastrum

G Gilia

Gy Gymnosteris

H Huthia

Ipo Ipomopsis

Lg Langloisia

Ln Linanthus

Lo Loeselia

Lp Leptodactylon

M Microsteris

N Navarretia

Ph Phlox

Po Polemonium

the concepts of phylogenetic relationships in the Phlox family which have emerged from the present study.

Separate phyletic lines within the family which have occupied the same or similar habitats have frequently converged in their morphological characters. Many instances of parallelisms which have obscured the systematic relationships are cited in Chapter 7. Parallel evolution is a resultant of the combined action of parallel genetic variation and parallel selection. The parallel variations can arise either through independent mutations at homologous loci in the different phylads, or through the sharing of the same mutations by separate phylads as a result of hybridization and gene exchange (Fig. 67). The parallel selection comes about when the different phyletic lines inhabit the same environment and are consequently exposed to some of the same selective factors. The occurrence of convergences between different lines within the Phlox family suggests strongly that the end results are products of natural selection, since no other mechanism is known which can bring about oriented evolutionary changes, and this in turn implies that the common characters of the separate phylads are adaptive.

The evolution of the Phlox family has taken place largely in an American setting. This evolution has run parallel to the major environmental changes which have occurred in the American hemisphere during the Tertiary and Quaternary periods. The ancestral Polemoniaceae, a plant with some of the characteristics of the modern genera *Cobaea, Cantua* and *Bonplandia*, is believed to have inhabited the neotropical rain-forest of early Tertiary time. During the Tertiary with the onset of more arid climatic conditions the northern and southern arms of the neotropical forest developed into a more xeric woodland vegetation. The evolutionary response of the Phlox family to these changes was the divergence of the Bonplandia tribe in the north and the Cantua tribe in the south from their common tropical ancestor.

As the climate continued to become increasingly arid during the Miocene and Pliocene, the subtropical woodland vegetation gave way in western North America to the more xeric phases of the Madro-Tertiary flora. The Gilia tribe is closely associated ecologically with the modern remnants of this flora and probably diverged from some ancestor belonging to the Bonplandia tribe as a response to the rise of more arid conditions.

The Polemonium tribe, meanwhile, underwent its major evolutionary development within the Arcto-Tertiary flora. The divergence of this tribe from the ancestral form probably represented a shift in ecological zone from an inhabitant of a tropical or subtropical rain-forest to an inhabitant of a temperate forest.

From the Pliocene through the Pleistocene to the Recent, as aridity and

cold came to dominate more and more regions of North America, the Polemoniaceae responded in characteristic fashion to the new conditions. The
Gilia tribe gave rise to many lines adapted to the deserts. The Polemonium
tribe colonized the alpine zones of the mountains, the arctic tundra, and
migrated through Siberia to western Europe. Both tribes sent representatives
across the tropics to various favorable areas in temperate South America.

The history of development of the Phlox family is not unique. Many families
and tribes of flowering plants are represented by groups with primitive
characteristics in the American tropics, by more advanced but still fairly
primitive groups in the Mexican highland, and by phylogenetically advanced
forms in western North America and the boreal region (Table 14). This suc-

Table 14. *The Parallel Development in Different Families of Genera
fitted for the Environmental Conditions in Four Subcontinental Regions*

Family or Tribe	American Tropics	Mexican Highland	Western North America	Boreal Region
Polemoniaceae	Cobaea	Bonplandia, Loeselia	Polemonieae, Gilieae	Polemonium, Phlox
Hydrophyllaceae	Wigandia	Nama	Phacelia, Nemophila	Romanzoffia
Boraginaceae	Cordioideae	Macromeria, Onosmodium	Allocarya, Plagiobotrys, Oreocarya, Cryptantha, Amsinckia	Mertensia
Solanaceae	Cestrum	Nicotiana	Nicotiana	
Scrophulariaceae Cheloneae	Dermatocalyx, Russelia, Uroskinnera, Berendtia, Tetranema	Penstemon	Penstemon	Scrophularia, Chelone
Ericaceae-Arbutoideae	Leucothoë	Comarostaphylis, Arbutus	Arctostaphylos, Arbutus	Arctostaphylos, Cassiope, Andromeda
Nyctaginaceae	Leucaster, Pisonia, Neea	Mirabilis, Allionia	Abronia	
Basellaceae and Portulacaceae	Basellaceae	Talinum, Talinopsis	Lewisia, Calyptridium, Spraguea, Montia	Montia, Claytonia
Onagraceae	Fuchsia, Hauya	Lopezia	Oenothera, Clarkia	Circaea, Epilobium
Cactaceae	Pereskia	Cereus	Opuntia	
Rosaceae-Dryadinae		Cowania, Fallugia	Cowania, Fallugia	Geum, Dryas

cession suggests an evolutionary progression from tropical ancestors through intermediate subtropical stages to derivative forms adapted to the more extreme environments now found in western North America and the far north. Such a progression is in line with the major climatic trend toward increasing aridity and cold during and since the Tertiary period.

In the colonization of temperate South America from North American centers the Polemoniaceae also stands in a numerous company. Examples of amphitropical distribution patterns belonging to the families Compositae, Cruciferae, Ericaceae, Gramineae, Hydrophyllaceae, Onagraceae, Rosaceae, Scrophulariaceae, Umbelliferae inter alia, have been briefly mentioned in the relevant discussion in Chapter 8. Unlike some other North American groups belonging to the Cactaceae, Onagraceae, Solanaceae and other families which have become dispersed to South America, the herbaceous Polemoniaceae have not developed an important secondary center of evolution in their new home. They constitute, on the contrary, an insignificant segment of the flora in the southern hemisphere.

The unaggressiveness of the austral Polemoniaceae is reflected in the fact that a whole genus or section of northern origin is represented by only one or two species in South America, which are either identical with or very closely related to particular North American species. The evolution of the temperate Polemoniaceae in South America appears to have been limited to the formation of some new ecotypes and one or two events of speciation for each phylad. This failure of the herbaceous Polemoniaceae to develop a secondary center of origin in South America may be due to the circumstance that the favorable habitats are few or were already occupied when the colonizers arrived.

The Phlox family has thus evolved through a series of stages from tropical shrubs or trees to ephemeral desert herbs and has migrated from one center of origin to another. Each stage served as a testing ground for the challenges of the future. Each stage prepared some lineages for the next stage by permitting them to become outfitted with the adaptive or "pre-adaptive" characteristics which they would need in the new habitats.

CLASSIFIED BIBLIOGRAPHY

EARLY HISTORY OF THE FAMILY

Taxonomic works since 1850 are listed in a separate bibliography under the heading of Taxonomy. Chapter 5 on Nomenclature contains some early references not included here.

ADANSON, M. 1763. Familles des plantes. 2 vol., Paris. (vol. 2, pp. 202–214)

BARTLING, F. G. 1830. Ordines naturales plantarum. Göttingen. (pp. 635–638).

BAUHIN, C. 1623. Pinax theatri botanici. Basel. (p. 164)

BAUHIN, J. and CHERLER, J. H. 1650–1651. Historia plantarum universalis. 3 vol., Yverdun, Switzerland. (vol. 3, p. 212)

BENTHAM, G. 1833. (Polemoniaceae). Bot. Reg. 19: t. 1622.

—. 1845. see Candolle and Candolle.

CANDOLLE, A. P. DE. 1813. Théorie élémentaire de la botanique. Paris. (pp. 213–220)

CANDOLLE, A. P. DE and CANDOLLE, A. DE. 1824–1870. Prodromus systematis naturalis regni vegetabilis. 17 vol., Paris. (vol. 9, pp. 302–322, 1845, by Bentham)

CANDOLLE, A. P. DE and LAMARCK, J. B. DE. 1805. Flore française. ed. 3, 4 vol., Paris. (vol. 3, pp. 645–646)

CAVANILLES, A. J. 1791. Icones et descriptiones plantarum. 6 vol., Madrid. (vol. 1, pp. 11–12)

—. 1800. Bonplandia. Anal. Hist. Nat. (Madrid) 2: 131–132.

DEWES, G. 1578. A nievve herball, or historie of plantes. London. (pp. 340–341)

DILLENIUS, J. J. 1732. Hortus elthamensis. 2 vol., London. (vol. 1, pp. 203–205)

DODONAEUS, R. 1559. De stirpium historia commentariorum imagines. Antwerp. (p. 344)

DON, D. 1822. Observations on the natural family of plants called Polemoniaceae; with descriptions of the genera belonging to it, and of a genus improperly referred to that order by botanists. Edinburgh Phil. Jour. 7: 283–291.

—. 1824. Observations on a new natural family of plants to be called Cobeaceae. Edinburgh Phil. Jour. 10: 109–112.

DON, G. 1838. A general history of the dichlamydeous plants. 4 vol., London. (vol. 4, pp. 236–249)

DUMORTIER, B. C. 1829. Analyse des familles des plantes. Tournay. (pp. 20, 25)

ENDLICHER, S. L. 1836–1840. Genera plantarum secundum ordines naturales disposita. Vienna. (pp. 656–658)

GAY, C. 1845–1853. Historia física y política de Chile. Botánica. 8 vol., Paris. (vol. 4, pp. 420–430, 1849)

GERARDE, J. 1633. The herball or generall historie of plantes. London. (p. 1076)

HOOKER, W. J. 1840. Flora boreali-americana. 2 vol., London. (vol. 2, pp. 71–76)

HOOKER, W. J. and ARNOTT, G. A. W. 1839. The botany of Captain Beechey's voyage. London. (pp. 364–369)

HUMBOLDT, A., BONPLAND, A. and KUNTH, C. S. 1815–1825. Nova genera et species plantarum. 7 vol., Paris. (vol. 3, pp. 160–166, 1818; vol. 6, pp. 83–85, 1823)

JUSSIEU, A. DE. 1789. Genera plantarum secundum ordines naturales disposita. Paris. (pp. 136–137)

—. 1804a. Mémoire sur le cantua, genre de plantes de la famille des Polémoniees. Ann. Mus. Hist. Nat. (Paris) 3: 113–119.

—. 1804b. Sur les caractères généraux de famille tirés des graines, et confirmés ou rectifiés par les observations de Gaertner. Ann. Mus. Hist. Nat. (Paris) 5: 246–265.

—. 1810. Sur les genres de plantes à ajouter ou retrancher aux familles des Solanées, Borraginées, Convolvulacées, Polemoniacées, Bignoniées, Gentianées, Apocinées, Sapotées et Ardisiaceés. Ann. Mus. Hist. Nat. (Paris) 15: 336–356.

LINDLEY, J. 1830. An introduction to the natural system of botany. London. (pp. 219–220)

—. 1846. The vegetable kingdom. ed. 1, London. (pp. 635–638)

LINNAEUS, C. 1737. Genera plantarum eorumque characteres naturales. ed. 1, Leiden. (pp. 46, 52, 348)

—. 1751. Philosophia botanica. ed. 1, Stockholm. (pp. 31, 35–36)

—. 1753. Species plantarum. 2 vol., Stockholm. (vol. 1, pp. 151–153, 162–163; vol. 2, p. 628)

MEISNER, C. F. 1836–1843. Plantarum vascularium genera secundum ordines naturales digesta. Leipzig. (p. 273, 1839)

MICHAUX, A. 1803. Flora boreali-americana. ed. 1, 2 vol., Paris. (vol. 1, pp. 141–145)

NUTTALL, TH. 1818. The genera of North American plants. 2 vol., Philadelphia. (vol. 1, pp. 124–127)

PARKINSON, J. 1640. Theatrum botanicum: the theater of plants. 2 vol., London. (p. 123)

PERSOON, C. 1805. Synopsis plantarum seu enchiridium botanicum. 2 vol., Paris and Tübingen. (vol. 1, pp. 185–187)

PLUKENET, L. 1692–1705. Opera omnia botanica. 6 vol., London. (vol. 3, t. 98, 1696; vol. 4, p. 233, 1696)

PURSH, F. T. 1814. Flora americae septentrionalis. 2 vol., London. (vol. 1, pp. 147–151)

REICHENBACH, H. G. L. 1837. Handbuch des natürlichen Pflanzensystems. ed. 1, Dresden and Leipzig. (p. 194)

ROEMER, J. J. and SCHULTES, J. A. 1817–1830. Caroli a Linné equitis systema vegetabilium. 7 vol., Stuttgart. (vol. 4, pp. 357–371, 1819)

RUIZ, L. and PAVON, J. 1794. Florae peruvianae et chilensis prodromus. ed. 1, Madrid. (pp. 20, 25–26, t. 4)

— and —. 1798–1802. Flora peruviana et chilensis. 4 vol., Madrid. (vol. 2, pp. 8, 17–18, tt. 123, 131–133, 1799)

SPACH, E. 1834–1848. Histoire naturelle des végétaux. Phanérogames. 14 vol., Paris. (vol. 9, pp. 106–125, 1840)

SPRENGEL, C. 1825–1828. Caroli Linnaei systema vegetabilium. 5 vol., Göttingen. (vol. 1, pp. 623–626, 1825)

TOURNEFORT, J. P. DE. 1700. Institutiones rei herberiae. 3 vol., Paris. (vol. 1, p. 146; vol. 2, t. 61)

VENTENAT, E. P. 1794. Tableau du règne végétal selon la méthode de Jussieu. 4 vol., Paris. (vol. 2, pp. 398–402)

WILLDENOW, K. L. 1797–1830. Caroli a Linné species plantarum. 6 vol., Berlin. (vol. 1, pp. 839–842, 878–879, 886–887, 1797; vol. 3, p. 323, 1800)

RELATIONSHIPS OF THE FAMILY

ABRAMS, L. 1951. Illustrated flora of the Pacific states. 3: 474. Stanford University Press.

ALEXNAT, W. 1922. Sero-diagnostische Untersuchungen über die Verwandtschaften innerhalb der Sympetalen. Bot. Arch. 1: 129–154.

BENSON, L. 1957. Plant classification. Heath and Co., Boston.

BESSEY, C. E. 1915. Phylogenetic taxonomy of flowering-plants. Ann. Missouri Bot. Gard. 2: 109–164.

BROWN, W. H. 1938. The bearing of nectaries on the phylogeny of flowering plants. Proc. Amer. Phil. Soc. 79: 549–594.

DAWSON, MARION L. 1936. The floral morphology of the Polemoniaceae. Amer. Jour. Bot. 23: 501–511.

DICKSON, JEAN. 1936. Studies in floral anatomy. III. An interpretation of the gynaeceum in the Primulaceae. Amer. Jour. Bot. 23: 385–393.

DOUGLAS, GERTRUDE E. 1936. Studies in the vascular anatomy of the Primulaceae. Amer. Jour. Bot. 23: 199–212.

ENGLER, A. 1891. (Preface to the Convolvulaceae). Engler and Prantl, Die natürlichen Pflanzenfamilien, 4 (3a): 1–3.

ENGLER, A. and GILG, E. 1924. Syllabus der Pflanzenfamilien. 10th ed., Berlin.

ERDTMAN, G. 1952. Pollen morphology and plant taxonomy. Angiosperms. Waltham, Mass.

FLORY, W. S. 1937. Chromosome numbers in the Polemoniaceae. Cytologia, Fujii Jub. Vol., 171–180.

GUNDERSEN, A. 1950. Families of dicotyledons. Waltham, Mass.

HALLIER, H. 1905. Provisional scheme of the natural, phylogenetic system of flowering plants. New Phytol. 4: 151–162.

HUMBOLDT, A., BONPLAND, A., and KUNTH, C. S. 1823. Nova genera et species plantarum. 6: 83–85.

HUTCHINSON, J. 1926. The families of flowering plants. I. Dicotyledons. London.

—. 1948. British flowering plants. London.

KAVALJIAN, L. G. 1952. The floral morphology of Clethra alnifolia with some notes on C. acuminata and C. arborea. Bot. Gaz. 113: 392–413.

LECHNER, SUSANNA. 1914. Anatomische Untersuchungen über die Gattungen Actinidia, Saurauia, Clethra und Clematoclethra mit besonderer Berücksichtigung ihrer Stellung im System. Beih. Bot. Centralbl. 32: 431–467.

MOSELEY, M. F. and BEEKS, R. M. 1955. Studies of the Garryaceae. I. The comparative morphology and phylogeny. Phytomorphology 5: 314–346.

RENDLE, A. B. 1925. The classification of flowering plants. II. Dicotyledons. Cambridge.

Soó, R. 1953. Die modernen Grundsätze der Phylogenie im neuen System der Blütenpflanzen. Acta Biol. Acad. Sci. Hungar. 4: 257–306.

SUNDAR RAO, Y. 1940. Male and female gametophytes of Polemonium coeruleum Linn., with a discussion on the affinities of the family Polemoniaceae. Proc. Nat. Inst. Sci. India 6: 695–704.

WERNHAM, H. F. 1911–1912. Floral evolution; with particular reference to the sympetalous dicotyledons. New Phytol. 10: 73–83, 109–120, 145–159, 217–226, 293–305. 11: 145–166, 217–235, 290–305, 373–397.

WETTSTEIN, R. 1935. Handbuch der systematischen Botanik. Leipzig & Vienna.

WHERRY, E. T. 1955. The genus Phlox. Morris Arboretum Monographs, III, Philadelphia.

MORPHOLOGY

BANCHER, E. 1953. Studien an der Blüte von Phlox panniculata hybr. Oesterr. Bot. Zeitschr. 100: 308–318.

BILLINGS, F. H. 1901. Beiträge zur Kenntnis der Samenentwicklung. Flora 88: 253–318. (Polemonium, Collomia, Phlox, Gilia, Linanthus)

BRAND, A. 1907. Polemoniaceae. A. Engler, Das Pflanzenreich, 4 (250): 2–13. (*Cobaea, Cantua, Polemonium, Allophyllum, Microsteris, Gilia, Ipomopsis, Navarretia, Linanthus*)

BROWN, W. H. 1938. The bearing of nectaries on the phylogeny of flowering plants. Proc. Amer. Phil. Soc. 79: 549–594. (*Phlox*)

COOPER, D. C. 1933. Nuclear divisions in the tapetal cells of certain angiosperms. Amer. Jour. Bot. 20: 358–364. (*Phlox*)

CRAMPTON, B. 1954. Morphological and ecological considerations in the classification of *Navarretia* (Polemoniaceae). Madroño 12: 225–238.

DAHLGREN, K. V. O. 1927. Die Morphologie des Nuzellus mit besonderer Berücksichtigung der deckzellosen Typen. Jahrb. Wiss. Bot. 67: 347–426. (*Cobaea*)

DAWSON, MARION L. 1936. The floral morphology of the Polemoniaceae. Amer. Jour. Bot. 23: 501–511. (*Cobaea, Cantua, Phlox, Navarretia, Linanthus*)

ERDTMAN, G. 1952. Pollen morphology and plant taxonomy. Angiosperms. Waltham, Mass. (*Cobaea, Cantua, Huthia, Bonplandia, Polemonium, Collomia, Gymnosteris, Phlox, Microsteris, Gilia, Langloisia, Navarretia, Linanthus*)

FARR, WANDA K. 1920. Cell divisions of the pollen-mother cell of *Cobaea scandens alba*. Bull. Torr. Bot. Club 47: 325–338.

GRANT, V. 1956. The genetic structure of races and species in *Gilia*. Adv. Genetics 8: 55–87.

JOHANSEN, D. A. 1950. Plant embryology. Chronica Botanica Co., Waltham, Mass. (*Polemonium, Phlox, Gilia, Linanthus*)

JUEL, H. O. 1915. Untersuchungen über die Auflösung der Tapetenzellen in den Pollensäcken der Angiospermen. Jahrb. Wiss. Bot. 56: 337–364. (*Cobaea, Polemonium*)

KRAEMER, H. 1910. The histology of the rhizome and roots of *Phlox ovata* L. (*Phlox carolina* L.). Amer. Jour. Pharm. 82: 470–475.

LAWSON, A. A. 1898. Some observations on the development of the karyokinetic spindle in the pollen-mother-cells of *Cobaea scandens* Cav. Proc. Calif. Acad. Sci., ser. 3, 1: 169–188.

MARTIN, A. C. 1946. The comparative internal morphology of seeds. Amer. Midl. Nat. 36: 513–660. (*Polemonium, Collomia, Phlox, Microsteris, Gilia, Ipomopsis, Navarretia, Linanthus*)

METCALFE, C. R. and CHALK, L. 1950. Anatomy of the Dicotyledons, 2: 939–943. (*Cobaea, Cantua, Loeselia, Polemonium, Collomia, Phlox, Gilia, Eriastrum*)

MILLER, HELENA A. and WETMORE, R. H. 1945a. Studies in the developmental anatomy of *Phlox Drummondii* Hook. I. The embryo. Amer. Jour. Bot. 32: 588–599.

— and —. 1945b. idem. II. The seedling. Amer. Jour. Bot. 32: 628–634.

— and —. 1946. idem III. The apices of the mature plant. Amer. Jour. Bot. 33: 1–10.

SCHNARF, K. 1921. Kleine Beiträge zur Entwicklungsgeschichte der Angiospermen. I. *Gilia millefoliata* Fisch. et Mey. Oesterr. Bot. Zeitschr. 70: 153–158.

—. 1929. Embryologie der Angiospermen. K. Linsbauer, Handbuch der Pflanzenanatomie. Abt. 2, Teil 2. (*Polemonium, Phlox, Gilia, Linanthus*)

—. 1937. Studien über den Bau der Pollenkörner der Angiospermen. Planta 27: 450–465. (*Phlox, Gilia*)

SOUÈGES, R. 1939a. Embryogénie des Polemoniacées. Développement de l'embryon chez le *Polemonium caeruleum* L. Compt. Rend. Acad. Sci. Paris 208: 1338–1340.

—. 1939b. Les lois du développement chez le *Polemonium caeruleum* L. Bull. Soc. Bot. France 86: 289–297.

—. 1942. Embryogénie des Polemoniacées. Développement de l'embryon chez le *Gilia tricolor* Benth. Compt. Rend. Acad. Sci. Paris 215: 543–545.

—. 1945. Embryogénie des Polemoniacées. Développement de l'embryon chez le *Polemonium pauciflorum* Wats. Compt. Rend. Acad. Sci. Paris 220: 897–900.

SUNDAR RAO, Y. 1940. Male and female gametophytes of *Polemonium coeruleum* Linn.,
with a discussion of the affinities of the family Polemoniaceae. Proc. Nat. Inst.
Sci. India 6: 695–704.

VUILLEMIN, P. 1909. L'héteromérie normale du *Phlox subulata*. Compt. Rend. Acad. Sci.
Paris 148: 650–652.

WAGNER, R. 1901. Über den Bau und die Aufblühfolge der Rispen von *Phlox paniculata*
L. Sitzungsber. kaiserl. Akad. Wissensch. Wien, Math.-Naturwiss. Classe, 110:
507–591.

TAXONOMY

Papers dealing solely with the descriptions of new species, range extensions, and
identification of type localities, as well as specialized papers superseded by later more
general treatments, are omitted from this bibliography. Floristic works containing
references to the Polemoniaceae are included in certain cases where the treatment represents a significant advance in the taxonomy of the family.

BAILLON, H. 1890. Polémoniacees. Histoire des Plantes, 10: 339–342.

BENTHAM, G. 1833. (Polemoniaceae) Bot. Reg. 19: t. 1622.

—. 1845. Polemoniaceae. A. de Candolle, Prodromus systematis naturalis regni
vegetabilis, 9: 302–322.

BENTHAM, G. and HOOKER, J. D. 1876. Polemoniaceae. Genera Plantarum, 2: 820–824.

BORSINI, OLGA E. 1942. Polemoniaceas argentinas. Lilloa 8: 199–230.

BRAND, A. 1907. Polemoniaceae. A. Engler, Das Pflanzenreich, 4 (250), 203 pp.

—. 1908. Polemoniacea peruviana. in, Plantae novae andinae imprimis Weberbaueria-
nae. IV, ed. I. Urban. Engler's Bot. Jhrb. 42: 174–175.

—. 1913a. Neue Beiträge zur Kenntnis der Polemoniaceen. Ann. Conserv. & Jard.
Bot. Genève 15/16: 322–342.

—. 1913b. Polemoniaceae peruvianae et bolivienses. Engler's Bot. Jahrb. 50, Beibl.
111: 50–52.

CONSTANCE, L. and ROLLINS, R. C. 1936. A revision of *Gilia congesta* and its allies. Amer.
Jour. Bot. 23: 433–440.

CRAIG, T. 1934. A revision of the subgenus *Hugelia* of the genus *Gilia* (Polemoniaceae).
Bull. Torrey Bot. Club 61: 385–396, 411–428.

CRAMPTON, B. 1954. Morphological and ecological considerations in the classification of
Navarretia (Polemoniaceae). Madroño 12: 225–238.

DAVIDSON, J. F. 1947. The present status of the genus *Polemoniella* Heller. Madroño 9:
58–60.

—. 1950. The genus *Polemonium* (Tournefort) L. Univ. Calif. Publ. Bot. 23: 209–282.

DON, D. 1824. Observations on a new natural family of plants to be called Cobeaceae.
Edinburgh Philosophical Jour. 10: 109–112.

EWAN, J. 1942. *Linanthastrum*, a new west American genus of Polemoniaceae. Jour.
Washington Acad. Sci. 32: 138–141.

GRANT, ALVA and GRANT, V. 1955. The genus *Allophyllum* (Polemoniaceae). El Aliso 3:
93–110.

GRANT, V. 1956. A synopsis of Ipomopsis. El Aliso 3: 351–362.

—. (and with Alva Grant). 1950–1956. Genetic and taxonomic studies in *Gilia*. I.-X.
El Aliso 2: 239–316, 361–388. 3: 1–91, 203–300.

—. 1957. The plant species in theory and practice. in, The species problem, ed. E. Mayr,
A. A. A. S. Symposium Volume.

GRAY, A. 1870. Revision of the North American Polemoniaceae. Proc. Amer. Acad. Sci., 8:
247–282.

—. 1878. Polemoniaceae. Synoptical Flora of North America, ed. 1, 2 (1): 128–151.

—. 1886. Polemoniaceae. idem, ed. 2, 2 (1): 128–151; Suppl. 406–412.

GREENE, E. L. 1887. Some American Polemoniaceae I. Pittonia 1: 120–139.

—. 1892. ibid. II. Pittonia 2: 251–260.

—. 1896. A new genus of Polemoniaceae. Pittonia 3: 29–30.

—. 1898. Some western Polemoniaceae. Pittonia 3: 299–305.

—. 1905. A proposed new genus, *Callisteris*. Leaflets of Botanical Observation and Criticism 1: 159–160.

HELLER, A. A. 1904. Western species, new and old. II. Muhlenbergia 1: 47–62.

—. 1906. Botanical exploration in California. Polemoniaceae. Muhlenbergia 2: 112–121.

—. 1912. The flora of the Ruby Mountains. V. Muhlenbergia 8: 49–58.

HEMSLEY, W. B. 1880. The Cobaeas. The Garden 17: 352–353.

HOUSE, H. D. 1908. The genus *Rosenbergia*. Muhlenbergia 4: 22–25.

HOWELL, J. T. 1938. A botanical visit to the Vancouver Pinnacles. Leafl. West Bot. 2: 97–102, 135–137.

JEPSON, W. L. 1925. Polemoniaceae. A Manual of the Flowering Plants of California, 781–809.

—. 1943. Polemoniaceae. A Flora of California, 3 (2): 131–222.

KEARNEY, T. H. and PEEBLES, R. H. 1943. *Gilia multiflora* Nutt. and its nearest relatives. Madroño 7: 59–63.

KINCH, W. H. 1956. A taxonomic study of *Microsteris* (Polemoniaceae). Unpubl. thesis, Claremont Graduate School, Claremont, Calif.

KLOKOV, M. V. 1955. Polemonia eurasiatica (in Russian). Botanical Materials from the Herbarium of the Komarov Institute 17: 273–323.

KUNTZE, O. 1891. Polemoniaceae. Revisio generum plantarum, 2: 432–434.

—. 1898. Polemoniaceae. idem, 3 (2): 202–203.

MACBRIDE, J. F. 1917. Notes on the Hydrophyllaceae and a few other North American spermatophytes. Contrib. Gray Herb. 49: 23–59.

MASON, H. L. 1941. The taxonomic status of *Microsteris* Greene. Madroño 6: 122–127.

—. 1945. The genus *Eriastrum* and the influence of Bentham and Gray upon the problem of generic confusion in Polemoniaceae. Madroño 8: 65–91.

—. 1950. Taxonomy, systematic botany and biosystematics. Madroño 10: 193–208.

—. 1951. Polemoniaceae. L. Abrams, Illustrated Flora of the Pacific States, 3: 396–474.

MASON, H. L. and GRANT, ALVA. 1948. Some problems in the genus *Gilia*. Madroño 9: 201–220.

— and —. 1951. *Gilia*. L. Abrams. Illustrated Flora of the Pacific States, 3: 456–474.

MEISNER, C. F. 1839. Polemoniaceae. Plantarum vascularium genera secundum ordines naturales digesta, 273.

MILLIKEN, JESSIE. 1904. A review of Californian Polemoniaceae. Univ. Calif. Publ. Bot. 2: 1–71.

MUNZ, P. A. 1957. Polemoniaceae. Munz and Keck, A California Flora, Claremont, in press.

NELSON, E. 1899. Revision of the western North American Phloxes. Wyoming Agr. Coll. Annual Report 9: 1–36.

PAYSON, E. B. 1924. *Collomia debilis* and its relatives. Univ. Wyo. Publ. Bot. 1: 79–87.

PETER, A. 1891. Polemoniaceae. Engler and Prantl, Die natürlichen Pflanzenfamilien, 4 (3a): 40–54.

REICHE, K. 1910. Polemoniaceas. Estudios Críticos sobre la Flora de Chile, 5: 146–158.

REICHENBACH, H. G. L. 1837. Polemoniariae. Handbuch des natürlichen Pflanzensystems, ed. 1, 194.

SHARSMITH, HELEN K. 1944. Notes on *Navarretia abramsii* of the Polemoniaceae. Amer.
    Midl. Nat. 32: 510–512.

STANDLEY, P. C. 1914. A revision of the genus *Cobaea*. Contrib. U. S. Natl. Herb. 17:
    448–458.

ST. JOHN, H. 1936. The replicate species of *Phlox* of the Pacific Northwest. Torreya 36:
    94–99.

VASSILJEV, V. N. 1953a. Notes on the systematics and geography of the genus *Polemonium*
    (in Russian). Botanical Materials from the Herbarium of the Komorov Institute
    15: 214–228.

—. 1953b. Polemoniaceae (in Russian). B. K. Shishkin et al, Flora U.S.S.R., 19: 77–95.

WEBERBAUER, A. 1945. El mundo vegetal de los Andes peruanos. Lima.

WHEELER, L. C. 1942. *Hugelia* Bentham preoccupied. Jour. Washington Acad. Sci. 32:
    237–239.

WHERRY, E. T. 1936. Miscellaneous eastern Polemoniaceae. Bartonia 18: 52–59.

—. 1940. A provisional key to the Polemoniaceae. Bartonia 20: 14–17.

—. 1942. The genus *Polemonium* in America. Amer. Midl. Nat. 27: 741–760.

—. 1943. *Microsteris*, *Phlox*, and an intermediate. Brittonia 5: 60–63.

—. 1944a. The minor genus *Polemoniella*. Amer. Midl. Nat. 31: 211–215.

—. 1944b. Review of the genera *Collomia* and *Gymnosteris*. Amer. Midl. Nat. 31:
    216–231.

—. 1945a. Supplementary notes on the genus *Polemonium*. Amer. Midl. Nat. 34:
    375–380.

—. 1945b. Two linanthoid genera. Amer. Midl. Nat. 34: 381–387.

—. 1946. The *Gilia aggregata* group. Bull. Torrey Bot. Club. 73: 194–202.

—. 1955. The genus *Phlox*. Morris Arboretum Monographs, 3, Philadelphia.

WHITEHOUSE, EULA. 1945. Annual *Phlox* species. Amer. Midl. Nat. 34: 388–401.

HORTICULTURE, GARDEN ESCAPES, ECONOMIC USES AND WEEDS

BACKER, C. A. 1951. Polemoniaceae. Flora Malesiana, ser. 1, 4 (3): 195–196.

BAILEY, L. H. (and with collaborators). 1933. *Cantua, Cobaea, Gilia, Polemonium, Phlox*.
    The Standard Cyclopedia of Horticulture, 1: 658, 806–807. 2: 1335–1337. 3: 2586–
    2591, 2729–2731.

BONSTEDT, C. 1932. Polemoniaceae. Pareys Blumengärtnerei, 2: 242–251.

BRAND, A. 1907. Polemoniaceae. A. Engler, Das Pflanzenreich 4 (250): 15–16.

CLAY, S. 1937. The present-day rock garden, 474–477.

EWART, A. J. 1930. Polemoniaceae. Flora of Victoria, 963–964.

GRANT, V. and Grant, ALVA. 1956. Genetic and taxonomic studies in *Gilia*. X. Conspec-
    tus of the subgenus *Gilia*. El Aliso 3: 297–300.

HEGI, G. 1927. Polemoniaceae. Illustrierte Flora von Mittel-Europa, 5(3): 2111–2119.

INGWERSEN, W. E. T. 1948. The genus *Phlox*. Ingwersen's Catalogues. East Grinstead,
    England.

RIDLEY, H. N. 1930. The dispersal of plants throughout the world. Ashford, Kent.

STANDLEY, P. C. 1924. Polemoniaceae. Trees and shrubs of Mexico. Contrib. U.S. Natl.
    Herb. 23: 1208–1213.

SYMONS-JEUNE, B. H. B. 1953. Phlox. Collins, London.

VASSILJEV, V. N. 1953. Polemoniaceae. B. K. Shishkin et al, Flora U.S.S.R., 19: 77–95.

WEBERBAUER, A. 1945. El mundo vegetal de los andes peruanos. Lima.

WHERRY, E. T. 1935. Our native Phloxes and their horticultural derivatives. Nat. Hort. Mag. 14: 209–231.

—. 1944. ·Ten American Polemoniums for the rock garden. Bull. Amer. Rock Garden Soc. 2: 8–11.

—·. 1946. Rock Garden Phloxes. Bull. Amer. Rock Garden Soc. 4: 17–27.

—. 1955. The genus *Phlox*. Morris Arboretum Monographs, III, Philadelphia.

CHROMOSOME NUMBERS

BEEKS, R. M. 1955. Improvements in the squash technique for plant chromosomes. El Aliso 3: 131–134.

CAIN, S. A. 1944. Foundations of plant geography. New York and London.

CLAUSEN, J. 1931. Genetic studies in *Polemonium*. III. Preliminary account on the cytology of species and specific hybrids. Hereditas 15: 62–66.

COVAS, G. and SCHNACK, B. 1946. Número de cromosomas en antófitas de la región de Cuyo (República Argentina). Rev. Argent. Agron. 13: 153–166.

DARLINGTON, C. D. 1937. Recent advances in cytology. 2nd. ed., London.

DARLINGTON, C. D. and WYLIE, A. P. 1955. Chromosome atlas of flowering plants. London.

DAVIDSON. J. F. 1950. The genus *Polemonium* (Tournefort) L. Univ. Calif. Publ. Bot. 23: 209–282.

FLORY, W. S. 1934. A cytological study on the genus *Phlox*. Cytologia 6: 1–18.

—. 1937. Chromosome numbers in the Polemoniaceae. Cytologia, Fujii Jub. ·Vol., 171–180.

FLOVIK, K. 1940. Chromosome numbers and polyploidy within the flora of Spitzbergen. Hereditas 26: 430–440.

GRANT, V. 1950. Genetic and taxonomic studies in *Gilia*. I. *Gilia capitata*. El Aliso 2: 239–316.

—. 1952a. idem. II. *Gilia capitata abrotanifolia*. El Aliso 2: 361–373.

—. 1952b. idem. III. The *Gilia tricolor* complex. El Aliso 2: 375–388.

—. 1952c. Cytogenetics of the hybrid *Gilia millefoliata* × *achilleaefolia*. Variations in meiosis and polyploidy rate as affected by nutritional and genetic conditions. Chromosoma 5: 372–390.

—. 1953. The role of hybridization in the evolution of the Leafy-stemmed Gilias. Evolution 7: 51–64.

—. 1954a. Genetic and taxonomic studies in *Gilia*. IV. *Gilia achilleaefolia*. El Aliso 3: 1–18.

—. 1954b. idem. V. *Gilia clivorum*. El Aliso 3: 19–34.

—. 1954c. idem. VI. Interspecific relationships in the Leafy-stemmed Gilias. El Aliso 3: 35–49.

—. 1956a. Chromosome repatterning and adaptation. Adv. in Genetics 8: 89–107.

—. 1956b. The influence of breeding habit on the outcome of natural hybridization in plants. Amer. Nat. 90: 319–322.

GRANT, V., BEEKS, R. M. and LATIMER, H. L. 1956. Genetic and taxonomic studies in *Gilia*. IX. Chromosome numbers in the Cobwebby Gilias. El Aliso 3: 289–296.

GRANT, V. and GRANT, ALVA. 1954. idem. VII. The Woodland Gilias. El Aliso 3: 59–91.

GRIESINGER, R. 1937. Ueber hypo- und hyperdiploide Formen von *Petunia*, *Hyoscyamus*, *Lamium* und einige andere Chromosomenzählungen. Ber. Deutsch. Bot. Ges. 55: 556–571.

KINCH, Wm. 1956. A systematic study of *Microsteris* (Polemoniaceae). Unpublished Thesis, Claremont, Calif.

LANGLET, O. 1936. Några bidrag till kännedomen om kromosomtalen inom Nymphaea-ceae, Ranunculaceae, Polemoniaceae och Compositae. Sv. Bot. Tidskr. 30: 288–294.

LÖVE, Á. and LÖVE, DORIS. 1948. Chromosome numbers of northern plant species. Dept. Agriculture, Reykjavik, Iceland.

MASON, H. L. 1941. The taxonomic status of *Microsteris* Greene. Madroño 6: 122–127.

McMILLAN, C. 1949. Documented chromosome numbers of plants. Madroño 10: 95.

MEYER, J. R. 1944. Chromosome studies of *Phlox*. Genetics 29: 199–216.

SUGIURA, T. 1936. Studies on the chromosome numbers in higher plants, with special reference to cytokinesis. I. Cytologia 7: 544–595.

—. 1940. idem. IV. Cytologia 10: 324–333.

TISCHLER, G. 1954. Das Problem der Basis-Chromosomenzahlen bei den Angiospermen-Gattungen und -Familien. Cytologia 19: 1–10.

ECOLOGY

BRAND, A. 1905. Kulturversuche mit verschiedenen Polemoniaceen-Arten. Engler's Bot. Jahrb. 36: 69–77. (germination of *Cobaea, Polemonium, Collomia, Gilia, Navarretia, Linanthus*)

CRAMPTON, B. 1954. Morphological and ecological considerations in the classification of *Navarretia* (Polemoniaceae). Madroño 12: 225–238. (habitats of *Navarretia*)

DAVIDSON, J. F. 1950. The genus *Polemonium* (Tournefort) L. Univ. Calif. Publ. Bot. 23: 209–282. (photoperiodism and growth responses of *Polemonium*)

EPLING, C. and DOBZHANSKY, TH. 1942. Genetics of natural populations. VI. Micro-geographic races in *Linanthus parryae*. Genetics 27: 317–332. (population dynamics of *Linanthus*)

GRANT, V. 1949. Seed germination in *Gilia capitata* and its relatives. Madroño 10: 87–93.

—. 1950. Genetic and taxonomic studies in *Gilia*. I. *Gilia capitata*. El Aliso 2: 239–316. (ecotypes in *Gilia*)

HORTON, J. S. and KRAEBEL, C. J. 1955. Development of vegetation after fire in the chamise chaparral of southern California. Ecology 36: 244–262. (successional ecology of *Allophyllum*)

HUNZIKER, J. H. 1952. La vegetación de la República Argentina. III. Las comunidades vegetales de la Cordillera de la Rioja. Revista Invest. Agríc. 6: 167–196. (habitats of *Gilia* and *Ipomopsis* in Argentina)

JUHREN, MARCELLA, WENT, F. W. and PHILLIPS, E. 1956. Ecology of desert plants. IV. Combined field and laboratory work on germination of annuals in the Joshua Tree National Monument, California. Ecology 37: 318–330. (germination of *Gilia, Erias-trum, Langloisia, Linanthus.*)

KRUCKEBERG, A. R. 1951. Intraspecific variability in the response of certain native plant species to serpentine soil. Amer. Jour. Bot. 38: 408–419. (ecotypes in *Gilia*)

LEWIS, H. and WENT, F. W. 1945. Plant growth under controlled conditions. IV. Res-ponse of California annuals to photoperiod and temperature. Amer. Jour. Bot. 32: 1–12. (physiological ecology of *Gilia* and *Linanthus*)

RIDLEY, H. N. 1930. The dispersal of plants throughout the world. Ashford, Kent. (dispersal mechanisms of *Polemonium, Collomia, Phlox, Gilia, Navarretia*)

SORIANO, A. 1950. La vegetación del Chubut. Revista Argent. Agron. 17: 30–66. (habitats of *Collomia* and *Microsteris* in Argentina)

SWEENEY, J. R. 1956. Responses of vegetation to fire. A study of the herbaceous vege-tation following chaparral fires. Univ. Calif. Publ. Bot. 28: 143–250. (germination of *Gilia* and *Allophyllum*; successional ecology of *Gilia, Allophyllum, Collomia, Linanthus, Navarretia*)

WEBERBAUER, A. 1945. El mundo vegetal de los Andes peruanos. Lima. (habitats of *Cobaea, Cantua, Huthia, Gilia* in Peru)

WENT, F. W. 1948. Ecology of desert plants. I. Observations on germination in the Joshua Tree National Monument, California. Ecology 29: 242–253. (germination of *Eriastrum* and *Linanthus*)

—. 1949. Ecology of desert plants. II. The effect of rain and temperature on germination and growth. Ecology 30: 1–13. (germination of *Gilia* and *Linanthus*)

WENT, F. W. and WESTERGAARD, M. 1949. Ecology of desert plants. III. Development of plants in the Death Valley National Monument. California. Ecology 30: 26–38. (habitats and germination of *Gilia* and *Langloisia*)

WRIGHT, S. 1943. An analysis of local variability of flower color in *Linanthus parryae*. (population dynamics of *Linanthus*)

PHYTOGEOGRAPHY

ABRAMS, L. 1915. The floral features of California. Popular Science Monthly 86: 22–30.

—. 1925. The origin and geographical affinities of the flora of California. Ecology 6: 1–6.

ATWOOD, W. W. 1940. The physiographic provinces of North America. New York.

AXELROD, D. I. 1938. The stratigraphic significance of a southern element in later Tertiary floras of western America. Jour. Wash. Acad. Sci. 28: 313–322.

—. 1939. A Miocene flora from the western border of the Mohave Desert. Carnegie Inst. Wash. Publ. 516.

—. 1940. Late Tertiary floras of the Great Basin and border areas. Bull. Torrey Bot. Club 67: 477–487.

—. 1950a. Classification of the Madro-Tertiary flora. Carnegie Inst. Wash. Publ. 590: 1–22.

—. 1950b. Evolution of desert vegetation in western North America. Carnegie Inst. Wash. Publ. 590: 215–306.

—. 1952. Variables affecting the probabilities of dispersal in geologic time. Bull. Amer. Mus. Nat. Hist. 99: 177–188.

BAKER, H. G. 1953. Dimorphism and monomorphism in the Plumbaginaceae. III. Correlation of geographical distribution patterns with dimorphism and monomorphism in *Limonium*. Ann. Bot. 17: 615–627.

—. 1955. Self-compatibility and establishment after "long-distance" dispersal. Evolution 9: 347–349.

BRAY, WM. L. 1898. On the relation of the flora of the Lower Sonoran Zone in North America to the flora of the arid zones of Chili and Argentine. Bot. Gaz. 26: 121–147.

—. 1900. The relations of the North American flora to that of South America. Science 12: 709–716.

CAIN, S. A. 1944. Foundations of plant geography. New York and London.

CAMPBELL, D. H. 1944. Relations of the temperate floras of North and South America. Proc. Calif. Acad. Sci. 25: 139–146.

CAMPBELL, D. H. and WIGGINS, I. L. 1947. Origins of the flora of California. Stanford Univ. Publ. Biol. Sci. 10: 1–20.

CHANEY, R. W. and MASON, H. L. 1936. A Pleistocene flora from Fairbanks, Alaska. Amer. Mus. Nov. 887.

CLEMENTS, F. E. 1920. Plant indicators. The relation of plant communities to process and practice. Carnegie Inst. Wash. Publ. 290.

COOKE, W. W. 1910. Distribution and migration of North American shorebirds. U. S. Dept. Agric., Biol. Survey. Bull. 35.

DAVIDSON, J. F. 1950. The genus *Polemonium* (Tournefort) L. Univ. Calif. Publ. Bot. 23: 209–282.

DEEVEY, E. S. 1949. Biogeography of the Pleistocene. Bull. Geol. Soc. Amer. 60: 1315–1416.

DuRietz, G. E. 1940. Problems of bipolar plant distribution. Acta Phytogeographica Suecica 13: 215–282.

Engler, A. 1879–1882. Versuch einer Entwicklungsgeschichte der Pflanzenwelt, insbesondere der Florengebiete seit der Tertiärperiode. 2 vol., Leipzig.

Godwin H. 1956. The history of the British flora. Cambridge.

Good, R. 1953. The geography of the flowering plants. 2nd ed., London.

Grant, V. 1950. Genetic and taxonomic studies in *Gilia*. I. *Gilia capitata*. El Aliso 2: 239–316.

Gray, A. 1884. Characteristics of the North American flora. Amer. Jour. Sci. & Arts, ser. 3, 28: 323–340. reprinted in C. S. Sargent, Scientific Papers of Asa Gray, 2: 260–282, 1889.

Gray, A. and Hooker, J. D. 1880. The vegetation of the Rocky Mountain region and a comparison with that of other parts of the world. Bull. U. S. Geol. and Geograph. Survey of the Territories 6: 1–77.

Harshberger, J. W. 1911. Phytogeographic survey of North America. A. Engler and O. Drude, Die Vegetation der Erde, vol. 13, Leipzig and New York.

Hemsley, W. B. 1886–1888. Botany. Appendix. F. D. Godman and O. Salvin, Biologia Centrali-Americana, Botany, 4: 117–315.

Hubbs, C. L. 1952. Antitropical distribution of fishes and other organisms. Proc. 7th Pacific Science Congr. 3: 324–329.

Jepson, W. L. 1923. A manual of the flowering plants of California. Berkeley.

Johnston, I. M. 1940. The floristic significance of shrubs common to North and South American deserts. Jour. Arnold Arb. 21: 356–363.

Kearney, T. H. and Peebles, R. H. 1942. Flowering plants and ferns of Arizona. Washington, D. C.

— and —. 1951. Arizona flora. Univ. California Press, Berkeley and Los Angeles.

Mason, H. L. 1942. Distributional history and fossil record of *Ceanothus*. M. van Rensselaer and H. E. McMinn, Ceanothus, 281–303. Santa Barbara, Calif.

Matthew, W. D. 1939. Climate and evolution. 2nd ed., New York Academy of Science, vol. 1.

Munz, P. A. 1935. A manual of southern California botany. San Francisco.

Penland, C. W. T. 1941. The alpine vegetation of the southern Rockies and the Ecuadorean Andes. Colorado College Publ. 230.

Pennell, F. W. 1935. The Scrophulariaceae of eastern temperate North America. Monographs of the Acad. Nat. Sci. Philadelphia, 1.

Reiche, K. 1907. Grundzüge der Pflanzenverbreitung in Chile. A. Engler and O. Drude, Die Vegetation der Erde, vol 8, Leipzig.

Shreve, F. 1942. The desert vegetation of North America. Bot. Rev. 8: 195–246.

Simpson, G. G. 1953. Evolution and geography. Oregon State System of Higher Education, Eugene.

Stebbins, G. L. 1950. Variation and evolution in plants. Columbia Univ. Press. New York.

Weber, Wm. A. 1955. Additions to the flora of Colorado. II. Univ. Colo. Studies, ser. Biology, 3 : 65–108.

Weberbauer, A. 1939. La influencia de cambios climáticos y geológicos sobre la flora de la costa peruana. Acad. Nac. Ciencias Exactas, Físicas y Naturales (Peru), 1939: 201–209.

—. 1945. El mundo vegetal de los Andes peruanos. 2nd ed., Lima.

Wherry, E. T. 1936. Miscellaneous eastern Polemoniaceae. Bartonia 18: 52–59.

—. 1940. Geographic relations in the genus *Phlox*. Bartonia 20: 12–14.

—. 1944a. The minor genus *Polemoniella*. Amer. Midl. Nat. 31: 211–215.

BIBLIOGRAPHY

—. 1944b. Review of the genera *Collomia* and *Gymnosteris*. Amer. Midl. Nat. 31: 216–231.

—. 1955. The genus *Phlox*. Morris Arboretum Monographs, 3, Philadelphia.

WULFF. E. V. 1943. An introduction to historical plant geography. transl., Chronica Botanica, vol. 10, Waltham, Mass.

PHYLOGENETIC PATTERNS (CHAPTERS 7 AND 9)

ANDERSON, E. and GAGE, Amy. 1952. Introgressive hybridization in *Phlox bifida*. Amer. Jour. Bot. 39: 399–404.

BRAUN, E. LUCY. 1956. Variation in *Polemonium reptans*. Rhodora 58: 103–116.

CHANEY, R. W. and MASON, H. L. 1936. A Pleistocene flora from Fairbanks, Alaska. Amer. Mus. Nov. 887.

DANSER, B. H. 1950. A theory of systematics. Bibliotheca Biotheoretica 4: 113–180.

GRANT, ALVA and GRANT, V. 1955. The genus *Allophyllum* (Polemoniaceae). El Aliso 3: 93–110.

— and —. 1956. Genetic and taxonomic studies in *Gilia* VIII. The Cobwebby Gilias. El Aliso 3: 203–287.

GRANT, V. 1950. Genetic and taxonomic studies in *Gilia*. I. *Gilia capitata*. El Aliso 2: 239–316.

—. 1953. The role of hybridization in the evolution of the Leafy-stemmed Gilias. Evolution 7: 51–64.

SIMPSON, G. G. 1953. The major features of evolution. Columbia Univ. Press, New York.

SPORNE, K. R. 1948. Correlation and classification in dicotyledons. Proc. Linn. Soc. London 160: 40–47.

—. 1956. The phylogenetic classification of the angiosperms. Biol. Rev. 31: 1–29.

STEBBINS, G. L. 1951. Push-button evolution. Quart. Rev. Biol. 26: 191–193.

INDEX

On pp. 130 ff. the reader will find an index to the names of species mentioned in the System of Classification (Chapter 4). The synonyms of these species are listed on these same pages. Generic and infrageneric names cited in Chapter 5 are indexed on pp. 126–127. Illustrations are indexed on pp. xiii – xiv.